I0510243

Hypovalent Cluster Structures

a programmed approach to solving problems using Wade's Rules

The Ten Educational Commandments

1. What you learn is controlled by what you already know and understand.

2. How you learn is controlled by how you have learned successfully in the past.

3. If learning is to be meaningful it has to link onto existing knowledge and skills, enriching and extending both.

4. The amount of material you can process at a time is limited.

5. Feedback and reassurance are necessary for comfortable learning, and assessment should be humane.

6. Individual approaches to learning and motivation should be considered.

7. Learning should be consolidated by asking yourself about what is going on in your own head.

8. You should have room for problem solving in its fullest sense to exercise and strengthen mental links.

9. You should have room to create, defend, try out, and hypothesise.

10. You should have chances to teach (you don't really learn until you teach).

Adapted from Johnstone *J. Chem. Ed.* 1997, **74(3)**, 262-268.

Other Programmed Chemistry Texts

I found this style of textbook useful when I was a student. Below is a list of all the programmed Chemistry books I am aware of, in case you want to make use of them.

1. *Chemistry of the Carbonyl Group* by Stuart Warren

2. *Organic Synthesis: The Disconnection Approach* by Stuart Warren

3. *Molecular Symmetry and Group Theory* by Alan Vincent

4. *Assigning Inorganic NMR Spectra* by Michael O'Neill

Table of Contents

Thanks

I have been supported in learning and teaching this topic by years of generous acts of collegiality from my teachers, my colleagues, and my students. Thank you all, friends.

I am particularly keen to thank those who helped with proofing an early version of this book. Alphabetically, these are: Sally Boss, Tim Morgan Boyd, Olie Townrow, and Caitilín McManus, and Alan Welch. All errors remaining are mine alone.

If, as a reader, you have any suggestions about edits I should make for the second edition, please contact me directly. This is quite an exposed model of publishing (self-publication through Amazon's print-on-demand platform), and I am keen to improve things iteratively while keeping textbook costs low for students.

MON, Oxford, April 2020

michaeloneill.org

A Small Dedication

My early University teaching was in the Supervision system at Cambridge. My PhD co-supervisor, Dr Sally Boss, lectured this topic and I gave Supervisions to small groups of students to follow up.

It was really noticeable how her lecturing got students excited about the topic. I believe that a large part of this was by helping them feel able to do the problems, rather than just using lecture time to transmit knowledge. Her lecturing somehow produced both emotional and cognitive effects.

Teaching resists measurement, but excellent teaching is often easy to recognise in the students it has changed. Seeing the student response to Sally's lecturing has been one of the largest positive influences in my own development as a teacher; making the same thing happen for my students is the goal I've pursued in my own teaching ever since.

Thanks, Sally.

How to Use This Book

Inspired by Warren's *Chemistry of the Carbonyl Group,* Vincent's *Molecular Symmetry and Group Theory*, and building on my experience of writing *Assigning Inorganic NMR Spectra*, I have structured this book as a 'programmed approach'. This format challenges you to solve problems through each chapter to construct your own understanding of the topic, and gives you feedback at every step. You can take things at your own pace. You can get things wrong before you get them right. No-one will ever know.

It won't teach you everything, but it will help you gain a working knowledge of the important points in a complex topic. You should see it as *one part* of your learning, alongside formal instruction and other textbooks. Some guidance on what to read next is given at the end of the last chapter.

You need to know a few things to use this book properly.

The whole book is split into *frames*. These are short sections separated by a page-wide horizontal line like this:

These lines are important because they map out a *sequence* of learning. You should physically cover up (with a scrap of paper or the cover of some lesser book) the frames you haven't reached yet.

You won't learn if you don't cover up the frames ahead of you; make good choices.

The text will occasionally have words in **bold**. **Bold** text is an instruction; you should do what the text says or answer the question it asks.

You must do this in writing rather than just in your head.

This will force you to create (actually, physically create) your own answers rather than just scrutinise mine. The process of failing is extremely valuable in learning; if you find out you were wrong, comparing the answers will help you see how you can improve your thinking.

Warren's advice still rings very true: "it is often only when you commit yourself to paper that you find out whether you really understand what you are doing".

I hope you find the programme enjoyable and helpful. If you find that this book format suits you, do look me up on Amazon or in your library – I'm trying to write a few of these, and the selection will hopefully grow over time.

MON, Oxford, April 2020

michaeloneill.org

A Note to Instructors

I don't think Wade's Rules should be a core part of undergraduate education. They are too specialist and *too perfect*: they are a beautiful, self-contained topic with a really nice set of examinable problems. In my opinion, our over-crowded degrees need to be addressing wicked problems much more extensively. Things like cancer and CO_2. Big things without tidy answers.

This book is therefore quite an interesting project for me. Finding the content unconvincing in justifying itself, I can see only four arguments for keeping Wade's Rules in a syllabus:

1. It is a local research specialism, and aligns with a research-led teaching agenda in an option course;
2. It is assessed in such a way that it somehow develops students' professional skills (e.g. they learn it independently or they give a spoken presentation on it);
3. It is knitted into a deep discussion of MO theory, developing students' understanding of core chemical principles;
4. It is aligned with some other core disciplinary skill (such as heteronuclear NMR or computational modelling) so that the structural problems are used as a platform for developing authentic disciplinary skills.

I have tried to heavily emphasise NMR assignment using the interludes between chapters, with resources in the Appendices to support a contact-time discussion of MO schemes. The programmed style should help students to develop confidence in their independent learning. The NMR focus should help develop their analytical skills.

I hope this allows the content to become a vehicle for developing some of the broader scientific skills which will help solve the knotty problems of this century.

MON, Oxford, April 2020, michaeloneill.org

Chapter 1: *Closo* Clusters

This book is about clusters: a class of molecules with a roughly-spherical arrangement of atoms. A representative cluster is $B_6H_6^{2-}$.

The core conceptual tool you will need to solve problems in this topic are a set of algorithms known as Wade's Rules. These rules allow you to use a cluster's formula (e.g. $B_6H_6^{2-}$) to predict its structure (an octahedron).

For the simple clusters in this chapter, Wade's Rules have three steps:

1. Identifying fragments;
2. Counting electrons; and
3. Selecting the structure from a list.

All fragments sit on the boundary between the inside of the cluster (endo) and the outside of the cluster (exo). In the example above, the B–H bond is exo to the cluster. The endo bonding is a complex molecular orbital scheme (Appendix I) which could be considered as six-centre bonding in the B_6 skeleton.

The B–H bond is a classical two-centre-two-electron bond. **Suggest how many electrons the BH fragment has 'left over'.**

Boron has three valence electrons, but uses one in forming a bond with hydrogen; it has two electrons 'left over'.

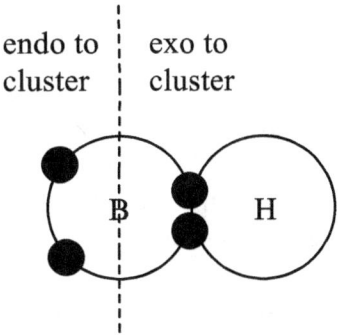

endo to cluster | exo to cluster

B   H

These two electrons are used in bonding the cluster skeleton (the B_6 octahedron) together. **How many electrons would a CH fragment donate to skeletal bonding?**

Carbon has four valence electrons, but uses one in exo C–H bonding; it donates three electrons to skeletal bonding.

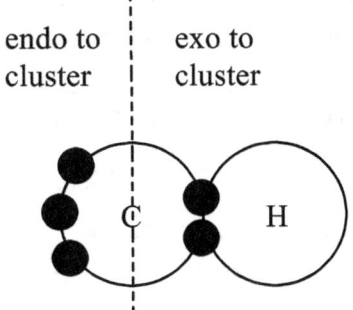

endo to cluster | exo to cluster

C   H

The charge of the cluster also contributes to the electron count within the skeleton. Each negative charge adds a count of one. **What is the total skeletal electron count in $B_6H_6^{2-}$?**

I tabulate my working when performing these calculations:

Fragment	Skeletal electrons per fragment	Number of fragments	Total skeletal electrons
HB	2	6	12
Charge	1	2	2

The total electron count is therefore 14.

Wade's Rules allow you to link the number 14 to the octahedral structure by following two steps:

1. Thinking of 14 electrons as 7 pairs of electrons; and
2. Subtracting one from seven, to give six.

This process will feel very uncomfortable to you at first. This book aims to get you using the rules before explaining why they work.

This 'six' tells you the number of corners on the shape of the cluster. The six-corner deltahedron is an octahedron. Appendix IV lists the shapes which correspond to the appropriate number of corners.

Looking back over this example, predict the shape of $C_2B_4H_6$. There are step-by-step instructions in the next few frames; look ahead if you are completely stuck, but try to attempt the problem first.

First, you need to identify the fragments. BH is one of the fragments. CH is the other. **How many electrons does each fragment contribute to the skeletal bonding?**

Each BH donates two electrons. Each CH donates three. **What is the total skeletal electron count?**

Fragment	Skeletal electrons per fragment	Number of fragments	Total skeletal electrons
HB	2	4	8
HC	3	2	6

How does the total count of 14 relate to a cluster shape?

Like the earlier example, the seven pairs of electrons suggest a six-corner shape (octahedral). Either isomer is fine at this stage (isomerism is discussed in Interlude 1).

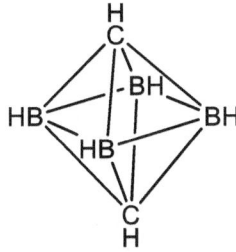

This algorithm is quite uncomfortable at first, but soon becomes familiar. The rest of this chapter is devoted to working through examples. You should feel very welcome to look back to help you master the process, and to use Appendix IV to look up shapes. Do not look ahead at the answers, though: correcting your own mistakes will help you to think better.

Predict the structure of $C_2B_3H_5$.

Fragment	Skeletal electrons per fragment	Number of fragments	Total skeletal electrons
HB	2	3	6
HC	3	2	6

Six pairs of electrons suggest a shape with five corners (trigonal bipyramid).

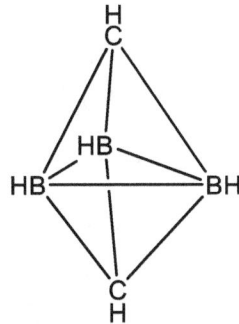

Predict the structure of $C_2B_5H_7$.

Fragment	Skeletal electrons per fragment	Number of fragments	Total skeletal electrons
HB	2	5	10
HC	3	2	6

The eight pairs of electrons suggest a seven-cornered shape (pentagonal bipyramid).

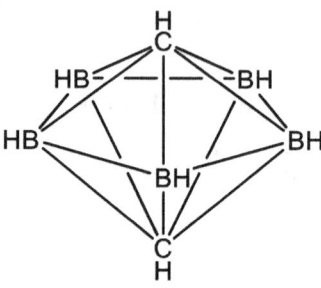

A harder one now. **Identify the fragments in B_5H_7.**

You have seen a few BH fragments now, but using five BH fragments still leaves you with two H atoms.

These H atoms are *not* incorporated into a BH_2 fragment. Instead, the hydrogen atoms sit in bridging positions between boron atoms. They are treated as contributing one electron each to the skeleton, but do not occupy a corner of the shape.

Perform the electron count and predict the shape.

Fragment	Skeletal electrons per fragment	Number of fragments	Total skeletal electrons
HB	2	5	10
Bridging H	1	2	2

The six pairs of electrons suggest a five-corner shape (trigonal bipyramidal). The 'bonds' drawn are not classical (Lewis) two-centre, two-electron bonds; instead, the lines only represent connectivity.

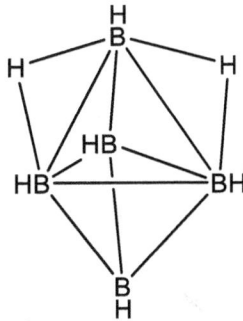

Predict the structure of CB_5H_7.

Fragment	Skeletal electrons per fragment	Number of fragments	Total skeletal electrons
HB	2	5	10
HC	3	1	3
Bridging H	1	1	1

The seven pairs of electrons suggests a six-corner shape (octahedral).

Another challenge. **Identify the fragments in B_9H_9S.**

You have already done a few examples involving the BH fragment. This leaves the S fragment.

All main group fragments have a pair of electrons exo to the cluster. This can be a bond (as you have seen in B–H and C–H fragments). **How might a 'naked' S fragment express a pair of exo electrons?**

15

S can present a *lone* pair of electrons exo to the cluster. With six valence electrons in total, this means that four electrons can be donated into the bonding skeleton.

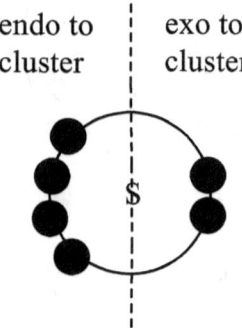

An exo lone pair is quite common behaviour in clusters incorporating heavier p-block elements. **Perform the electron count for B_9H_9S and predict the structure.**

Fragment	Skeletal electrons per fragment	Number of fragments	Total skeletal electrons
HB	2	9	18
:S	4	1	4

The eleven pairs of electrons in the bonding skeleton suggest that the structure should be based on a ten-corner shape. This is an unfriendly shape, but can be thought about as a bicapped square antiprism.

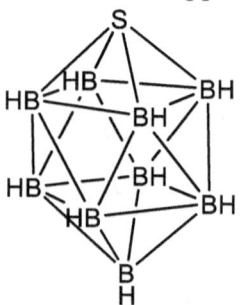

In general, I always advise students to describe (with words) the shape they are drawing because it is very easy to draw badly. Here, for example, you could write 10-CORNER SHAPE.

16

Predict the structure of $Sn(CR)_2B_4H_4$.

Fragment	Skeletal electrons per fragment	Number of fragments	Total skeletal electrons
HB	2	4	8
RC	3	2	6
:Sn	2	1	2

The eight pairs of electrons suggest that the cluster should adopt a seven-corner shape (pentagonal bipyramid).

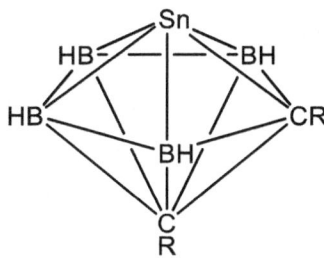

Predict the structure of Pb_5^{2-}.

This unusual specimen is an example of a 'Zintl' ion: an anion made up entirely of main group metal atoms. Remembering that every lead atom will have a lone pair exo to the bonding skeleton:

Fragment	Skeletal electrons per fragment	Number of fragments	Total skeletal electrons
:Pb	2	5	10
Charge	1	2	2

The six pairs of electrons suggest a five-corner shape (trigonal bipyramid).

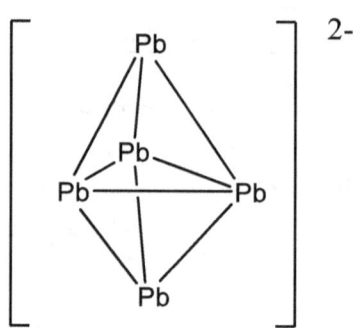

Predict the structure of TlSn₈³⁻.

$$\text{Predict the structure of } TlSn_8^{3-}.$$

Fragment	Skeletal electrons per fragment	Number of fragments	Total skeletal electrons
:Tl	1	1	1
:Sn	2	8	16
Charge	1	3	3

The ten pairs of electrons suggest a cluster shape with nine corners (tricapped trigonal prism).

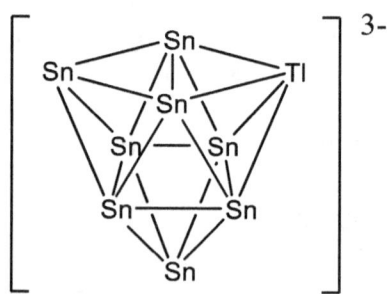

This concludes Chapter 1. While you might have some reservations about exactly *why* they work, you should be becoming confident in using Wade's Rules to rationalise the structure of main group clusters with a variety of shapes (octahedral, trigonal bipyramidal etc). You will likely be fairly comfortable with identifying the fragments and charges which contribute to the skeletal electron count, and starting to get the hang of counting the pairs and then subtracting one to determine the shape.

If you are still struggling, this is a good chance to go back and work through parts you found difficult. Mastering the nuts and bolts at this point is an excellent use of your time because the core skill of counting electrons just gets applied to progressively weirder structures.

Interlude 1: Carborane Isomers

Predict the structure of $C_2B_4H_6$.

First identify the fragments.

There are two types of fragment: CH and BH. **How many electrons does each of these contribute to the cluster?**

BH contributes two electrons; CH contributes three.

Fragment	Skeletal electrons per fragment	Number of fragments	Total skeletal electrons
HB	2	4	8
HC	3	2	6

What platonic shape does this electron count predict?

14 electrons is seven pairs. Seven pairs is characteristic of a six-corner shape (octahedron). **Draw the product.**

This is where it gets interesting: there are *two* possible products.

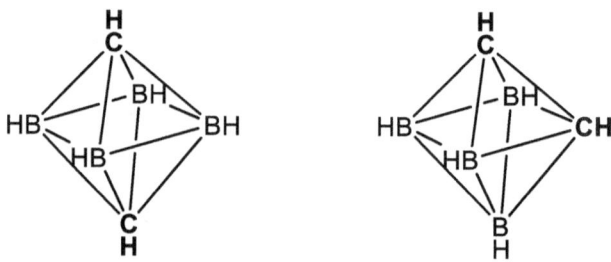

The 1,2-isomer has adjacent carbon atoms, the 1,6-isomer has no direct C–C interactions.

Both of these satisfy Wade's Rules, but geometric isomers generally have different energies; one of them will be more stable than the other. To determine the thermodynamic structure, you need to think about the bonds. **Which of the following bonds has a dipole?**

$$B–B \qquad B–C \qquad C–C$$

Bonds between the same element do not (to a first approximation) develop a dipole; only B–C has a dipole.

How might polarity help rationalise the placement of the carbon atoms?

The most widely-accepted argument is that the partial negative charge on carbon atoms leads to carbon-carbon repulsion, weakening the C–C bond. The number of C–C bonds is therefore minimised in the thermodynamic structure.

By the same token, B–B bonds are weakened by the repulsion between atoms with partial positive charge. The number of B–B bonds is also minimised in the thermodynamic structure.

Which $C_2B_4H_6$ isomer is most stable?

The 1,6-isomer is most stable because it minimises the number of C–C and B–B bonds (which here is equivalent to maximising the number of B–C bonds).

Predict the structure of $C_2B_3H_5$.

Again, you should identify and analyse each fragment:

Fragment	Skeletal electrons per fragment	Number of fragments	Total skeletal electrons
HB	2	3	6
HC	3	2	6

The twelve electrons suggest that the parent shape should have five corners (trigonal bipyramid). **Draw the structure(s) of the cluster.**

Again, there are three possible isomers: $1,2$-$C_2B_3H_5$, $2,3$-$C_2B_3H_5$, and $1,5$-$C_2B_3H_5$. **Which of these is the thermodynamic product?** (Note: Appendix IV has the numbering scheme for positions on a trigonal bipyramid.)

The 1,5-isomer minimises the number of homoatomic bonds, and so is most stable. Note that this is also consistent with another consideration: size. The smaller C atom is linked to three other atoms in the cluster, while the larger B atom connects to four atoms.

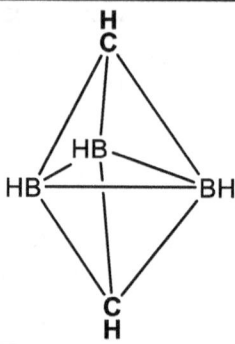

Interestingly, the cis-isomers are often the *kinetic* product in carborane synthesis as ethyne derivatives are commonly used to make them. The adjacent carbon atoms are preserved from the starting material through into the product because the C–C bond is energetically hard to break (due to a high activation energy) even if it is thermodynamically favourable to do so (to minimise the coulombic repulsion between partial negative charges).

The classic way of distinguishing isomers experimentally is by NMR spectroscopy: the number of carbon or boron environments is typically characteristic of the isomer. Later interludes will deal with using $^{13}C\{^1H\}$ and $^{11}B\{^1H\}$ NMR spectroscopy to characterise isomers.

The anion $B_5H_5^{4-}$ has a shape consistent with Wade's Rules. **Count the skeletal electrons in $B_5H_5^{4-}$.**

Fragment	Skeletal electrons per fragment	Number of fragments	Total skeletal electrons
HB	2	5	10
Charge	1	4	4

What parent shape does this electron count match?

The seven pairs predicts a six-corner (octahedral) shape. **Why is this a problem?**

There aren't six fragments to place on the six corners. The structure of $B_5H_5^{4-}$ is shown below. **How does it relate to an octahedron?**

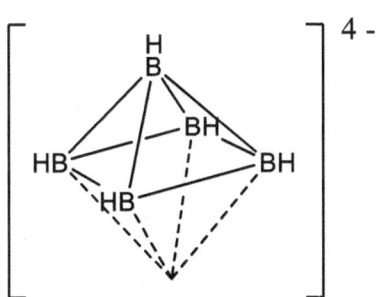

The parent octahedron is missing one corner, 'leaving behind' a square pyramid. This is a case of a general class of cluster described as '*nido*' (NIGH-doe) from the Greek word for 'nest'. We say that $B_5H_5^{4-}$ has a *nido*-octahedral shape, or is a *nido* octahedron. There is nothing 'at' the missing point: no atoms, no electrons. Nothing.

The structures you saw in Chapter 1 were all '*closo*': the parent shape (octahedron, trigonal bipyramid, etc) had a chemical fragment on every corner, so it matched the parent shape exactly. '*Closo*' comes from the Latin word for 'cage'.

Predict the structure of P_4 using Wade's Rules.

Fragment	Skeletal electrons per fragment	Number of fragments	Total skeletal electrons
:P	3	4	12

Six pairs of electrons suggest that the parent shape should have five corners (trigonal bipyramid), but there are only four atoms. The structure is a *nido* trigonal bipyramid.

This seems a little artificial – the molecule presents itself as a tetrahedron for all intents and purposes – but the structure is also consistent with Wade's Rules.

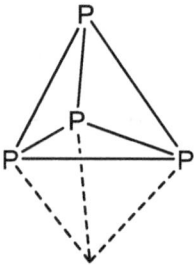

P_4 is a slightly special case because it actually has exactly the right electron count to form two centre, two electron P–P bonds. Wade's Rules only strictly hold for clusters with fewer electrons than this ('electron deficient' clusters); P_4 is on the very edges of this bonding model as it is 'electron precise': the structure can instead be described with classical Lewis bonds. Arguably, it is just a coincidence that the two bonding models agree for P_4.

$C_3B_3H_7$ has a *nido* pentagonal bipyramidal shape. **Show that this is consistent with Wade's Rules.**

Identifying the fragments here is a little tricky – remember that bridging hydrogen atoms each contribute one electron to the skeleton.

Fragment	Skeletal electrons per fragment	Number of fragments	Total skeletal electrons
HB	2	3	6
HC	3	3	9
Bridging H	1	1	1

The eight pairs of electrons suggest a parent shape with seven corners (pentagonal bipyramid), but there are only six terminal fragments. This gives a *nido* structure.

Looking at the experimental structure (note how the smaller carbon occupies the apical site, suggesting that this is not the thermodynamic structure), we can learn a couple more things from this particular example.

First, the point which is 'lost' when the parent shape presents in a *nido* cluster is the apex of the pyramid rather than the base (the equator of the bipyramid). In general (i.e. with other parent shapes), the lost point is the most connected one: here, the point connected to five atoms is removed rather than one connected to four.

Second, the bridging hydrogen atom arranges itself around the now-open face. It is hard to predict exactly where.

Predict the shape of $C_2B_9H_{11}^-$.

Fragment	Skeletal electrons per fragment	Number of fragments	Total skeletal electrons
HC	3	2	6
HB	2	9	18
Bridging H	1	4	1
Negative charge	1	1	1

Thirteen electron pairs suggest that the parent shape should have twelve corners (Appendix IV shows the shape for this number of corners). There are only eleven fragments, so this structure must be *nido*.

The apical position is lost when arranging atoms, and bridging hydride is arranged around the now-open face. Experimentally, this open face contains both C atoms; note that in the thermodynamic structure we would expect one of these to sit on the remaining apex.

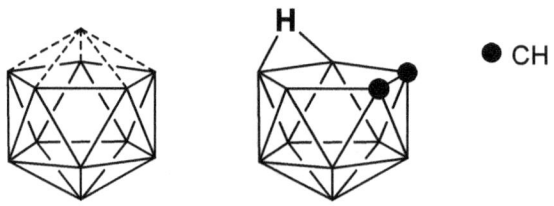

Predict the shape of B_5H_{11}.

Fragment	Skeletal electrons per fragment	Number of fragments	Total skeletal electrons
HB	2	5	10
Bridging H	1	6	6

Eight pairs of electrons suggests that the parent shape has seven corners (pentagonal bipyramid). **Suggest a structure based on a pentagonal bipyramid.**

The apex point is 'lost' first just like in the *nido* cluster (the point with highest connectivity), but then you could in principle lose either the other apex (giving a pentagonal planar molecule) or a point from the base of the pyramid. In practice, the loss happens on the base of the pyramid.

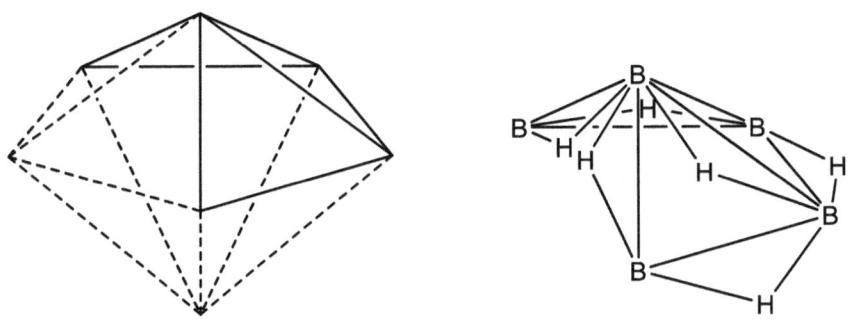

This is a specific case of a general rule. The second loss is *next to* the first one (and the third loss would be next to *both* previous lost points).

This class of molecule is called *arachno* (ah-RACK-no) to distinguish it from *closo* and *nido*. *Arachno* comes from the Greek word for 'web' or 'spider' or something.

There are cases where clusters lose even more points from their parent shapes, but they are rare and this book won't deal with them: *arachno* is as far as we'll go.

Predict the structure of P₄R₂.

Identifying the fragments is a bit difficult here. There are two :P fragments, with a lone pair exo to the cluster; and there are two R–P fragments which donate four electrons to the cluster (one of phosphorus' electrons is involved in P–C bonding.

Fragment	Skeletal electrons per fragment	Number of fragments	Total skeletal electrons
:P	3	2	6
RP	4	2	8

27

The seven pairs of electrons suggest that the parent shape should have six corners (octahedron). There are only four atoms available to decorate the points of the octahedron, so this cluster is *arachno*.

Losing the first point is easy because the octahedral symmetry means that all sites are equivalent. The second loss 'peels open' the cluster by locating next to the first absence, leaving a distinctive shape sometimes described as a butterfly.

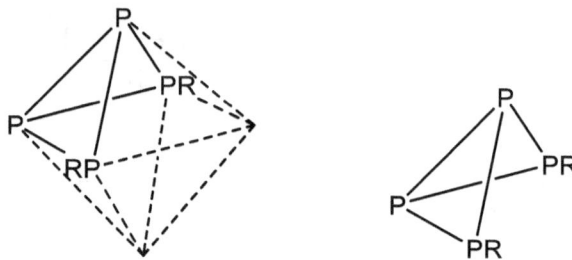

Wade's Rules do not suggest where the R groups should go (on the wings or the body of the butterfly); it would be unreasonable to ask this of you in an exam without further information (e.g. NMR data).

Predict the structure of $B_{10}H_{12}^{4-}$.

Fragment	Skeletal electrons per fragment	Number of fragments	Total skeletal electrons
HB	2	10	20
Bridging H	1	2	2
Charge	1	4	4

The thirteen pairs of electrons suggest a parent shape with 12 corners. Only 10 groups are available, so the cluster is *arachno*.

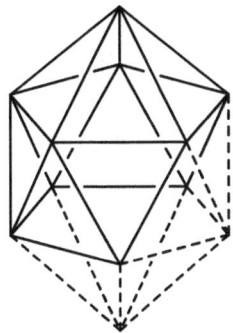 

It is hard to draw big shapes like this under exam conditions, and I have left the BH corners unlabelled for clarity. I personally find it helpful to think of an icosahedron as a bicapped pentagonal antiprism, but I strongly recommend using words to describe your shape in an exam. Your drawing might *look* like a box of broken pencils, but the words ARACHNO BASED ON 12-CORNER SHAPE will help the examiner to credit your scientific reasoning.

A harder example to finish the chapter on. **Predict the structure of $B_5H_8(PMe_2)$.**

The unusual fragment here is PMe_2. As shown in Chapter 1, you can use a 'dot' model to work out how the phosphorus electrons contribute to the cluster:

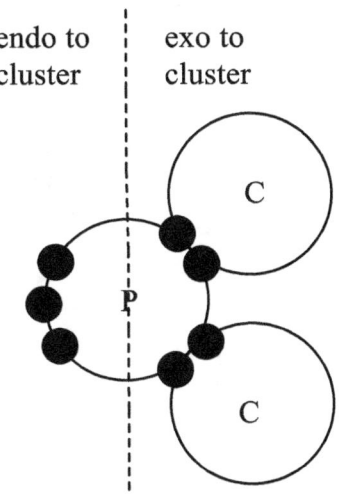

Fragment	Skeletal electrons per fragment	Number of fragments	Total skeletal electrons
HB	2	5	10
PMe$_2$	3	1	3
Bridging H	1	3	3

Suggest a shape for the cluster.

Experimentally, the phosphorus atom sits in the basal position of a *nido* pentagonal bipyramid, though you might have expected the larger atom to occupy the more-connected apical site.

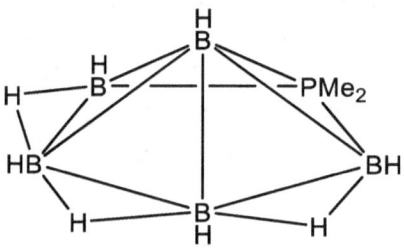

This concludes the chapter. By now you should be starting to get the hang of applying Wade's Rules to a range of main group clusters. Your understanding of the parent shape algorithm should be quite strong, and you should be starting to get a feel for how the loss of sites in *nido* and *arachno* clusters plays out. This is a good point to revisit any ideas you feel uncomfortable with so far, perhaps by working through some frames again or consulting another source like a different textbook or your lecture notes (in particular, many books adopt a different approach to working out the *closo*/*nido*/*arachno* state of a cluster).

Interlude 2: ^{13}C NMR Spectroscopy

Wade's Rules predict the most stable structure of an electron-deficient cluster, but this isn't always the observed product. Kinetic structures happen, by-products happen, cluster degradation happens. Even in a completely reliable reaction, chemists should be verifying that they have made what they set out to make.

The most convenient verification tool in carborane clusters is $^{13}C\{^{1}H\}$ NMR spectroscopy. It's relatively quick, it's very easy to perform (most NMR machines are set up to run carbon spectra anyway because organic chemists use it so often), and the interpretation is clear.

The $^{13}C\{^{1}H\}$ NMR spectrum of a sample of $C_2B_4H_8$ contains two peaks. **Suggest a structure for this cluster.**

Interpreting spectra is really hard if you don't have an idea about what a reasonable structure looks like. I solve assignment problems by suggesting a structure, seeing if it fits, and then modifying the structure if I need to. You may have a different approach; I'm going to work through mine by starting with Wade's Rules.

Fragment	Skeletal electrons per fragment	Number of fragments	Total skeletal electrons
HB	2	4	8
HC	3	2	6
Bridging H	1	2	2

The eight pairs of electrons suggest that the parent shape should have seven corners (pentagonal bipyramid). There are only six terminal groups, so the cluster is *nido*. Spacing the CH fragments out to maximise the number of strong polar C–B interactions gives the thermodynamic 2,4-product:

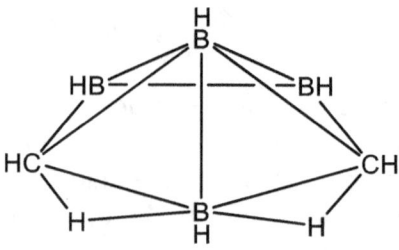

Does this fit the spectrum?

The carbon atoms are related by a symmetry operation (they can reflect onto each other), so they are equivalent. This should mean that each atom gives the same chemical shift, leading to a spectrum with one peak. The spectrum shows two peaks, so this structure must be wrong.

How could you modify the structure to try again?

The carbons need to be in different environments, so we could try the 1,2-isomer.

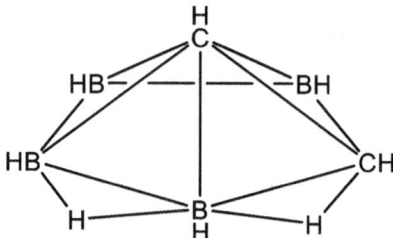

Does this fit the spectrum?

Yes. The apical site is unique in the structure, so the shift of an apical carbon atom will also be unique. The spectrum reflects one apical and one basal carbon. Any other isomer (e.g. placing the CH fragments at different basal sites) would result in only one ^{13}C environment.

The $^{13}C\{^1H\}$ NMR spectrum of a sample of $C_4B_2H_6$ contains three peaks. **Suggest a structure for this cluster.**

Again, I start by using Wade's Rules to propose a structure.

Fragment	Skeletal electrons per fragment	Number of fragments	Total skeletal electrons
HB	2	2	4
HC	3	4	12

The eight pairs of electrons suggest a parent shape with seven corners (pentagonal bipyramid). There are only six terminal groups, so the cluster must be *nido*. BH is in the most connected location (the apex of the pyramid).

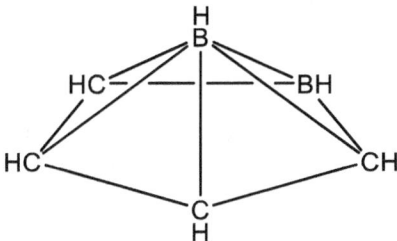

Does this fit the spectrum?

No. There are two carbon environments in the proposed structure, but the spectrum shows three environments. **Propose another structure.**

I have tried placing the BH fragments next to each other in the basal plane. This produces three carbon environments (as the 1,2,3,4-isomer has equivalent 2- and 4-positions) which is consistent with the spectrum.

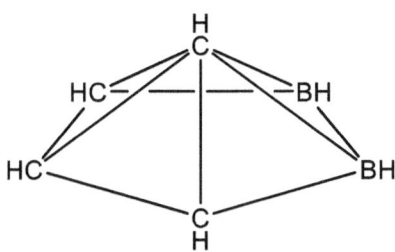

An interesting thing to note in closing is that the final possible isomer (with BH in the 2,4-positions) *also* fits the spectrum:

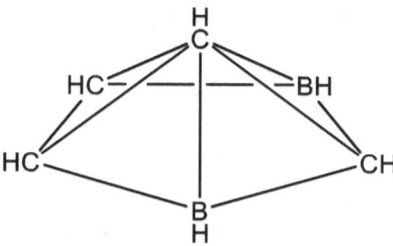

Without being familiar with the shifts commonly observed in carborane characterisation, it is impossible to distinguish which of these isomers is in the sample. This sort of ambiguity is surprisingly common in the highly symmetrical molecules which inorganic chemists seem to make.

For those of you interested in exploring Inorganic NMR spectra more deeply, I have written a very reasonably-priced book on just that topic: *Assigning Inorganic NMR Spectra*. The 'programmed' format of this book is also used in that one, too.

The next interlude will look at the use of $^{11}B\{^{1}H\}$ NMR spectroscopy in structural determination of carboranes.

Chapter 3: Clusters involving Transition Metals

The cluster compound $[Pb_9Cr(CO)_3]^{4-}$ has a *closo* decahedral (10-corner) structure. **How many electrons does the $Cr(CO)_3$ fragment contribute to the skeletal electron count?**

I like to solve these 'backwards' problems using algebra, but some people use a more intuitive approach. I am going to call the unknown electron count x.

Fragment	Skeletal electrons per fragment	Number of fragments	Total skeletal electrons
Pb	2	9	18
$Cr(CO)_3$	x	1	x
Charge	1	4	4

This gives $11 + \frac{x}{2}$ pairs of electrons.

If the shape is based on a 10-corner shape, it must have 11 pairs of electrons. This suggests that x is zero: the chromium fragment provides no electrons for skeletal bonding.

There is a general way of working out how many electrons a transition metal fragment:

$$electron\ count = group\ number + ligand\ electrons - 12$$

This is commonly represented as:

$$M = V + L - 12$$

We can analyse the $Cr(CO)_3$ fragment to help understand what these terms mean.

Chromium is in group 6, so V (the valence electron count) is 6. Formally, this treats chromium as being in the zero oxidation state because we deal with the overall charge separately in the Wade's Rules count for the overall cluster. If you have studied organometallic electron counting schemes, this is the basis of the neutral scheme.

The three carbonyl ligands – like in organometallic counting schemes – each contribute two electrons. L is therefore 6.

The neutral organometallic counting scheme helps to systematise ligand counts: PR_3 ligands also donate two electrons, but ligands treated as one-electron donors include CN, halogens, and alkyls.

Calculate M for the $Cr(CO)_3$ fragment.

Twelve is just a constant, so we can now plug all the values in:

$$M = V + L - 12$$

becomes

$$M = 6 + 6 - 12 = 0$$

Which is what our analysis of the cluster told us to begin with.

Using Wade's Rules, predict the structure of $C_2R_2B_4H_4Ru(\eta^6\text{-}C_6H_6)$. Take your time and look back over the chromium example if you need to.

The group 8 metal ruthenium receives 6 electrons from the benzene ligand:

$$M = V + L - 12$$

becomes

$$M = 8 + 6 - 12 = 2$$

The cluster electron count then becomes

Fragment	Skeletal electrons per fragment	Number of fragments	Total skeletal electrons
BH	2	4	8
CR	3	2	6
$Ru(\eta^6\text{-}C_6H_6)$	2	1	2

16 electrons is 8 pairs, which suggests a parent shape with 7 corners (pentagonal bipyramid). There are 7 fragments, so the cluster is *closo*. Experimentally, the ruthenium sits at the apical site. Note that this is consistent with arguments based on size: the large Ru atom is able to bond to the five basal sites more readily than the smaller B and C atoms.

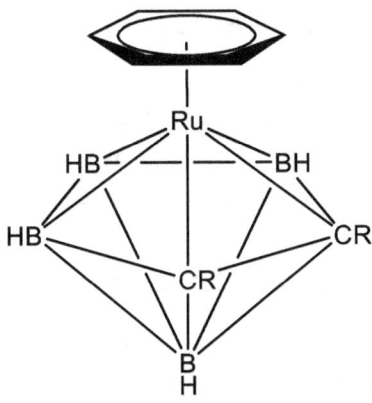

Predict the structure of $Co(\eta^5\text{-}C_5H_5)B_4H_8$.

Group 9 cobalt is bound to the five-electron cyclopentadienyl ligand, so:

$$M = V + L - 12$$

becomes

$$M = 9 + 5 - 12 = 2$$

Fragment	Skeletal electrons per fragment	Number of fragments	Total skeletal electrons
BH	2	4	8
Bridging H	1	4	4
$Co(\eta^5\text{-}C_5H_5)$	2	1	2

The seven pairs of electrons suggest that the parent shape should have six corners (octahedron). There are only five terminal fragments, so the cluster is *nido*.

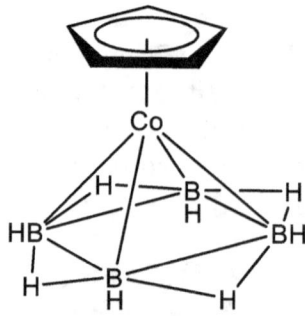

Predict the structure of $Fe(CO)_3(C_2B_3H_7)$.

Iron is in group 8 and has three two-electron ligands, so
$$M = V + L - 12$$
becomes
$$M = 8 + 6 - 12 = 2$$

Fragment	Skeletal electrons per fragment	Number of fragments	Total skeletal electrons
BH	2	3	6
CH	3	2	6
Bridging H	1	2	2
$Fe(CO)_3$	2	1	2

The parent 7-corner shape (trigonal bipyramid) presents as a *nido* cluster.

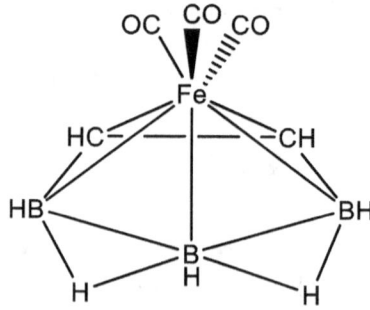

Predict the structure of $(Ph_3P)_2HRh(C_2B_9H_{11})$.

Rhodium is in group 9. It has two two-electron phosphine ligands, and a one-electron hydride (neutral counting).

$$M = V + L - 12$$

becomes

$$M = 9 + 5 - 12 = 2$$

Fragment	Skeletal electrons per fragment	Number of fragments	Total skeletal electrons
BH	2	9	18
CH	3	2	6
Rh(PPh₃)₂H	2	1	2

Thirteen pairs of electrons suggest a parent 12-corner shape. As there are also twelve terminal fragments, this presents as a *closo* cluster.

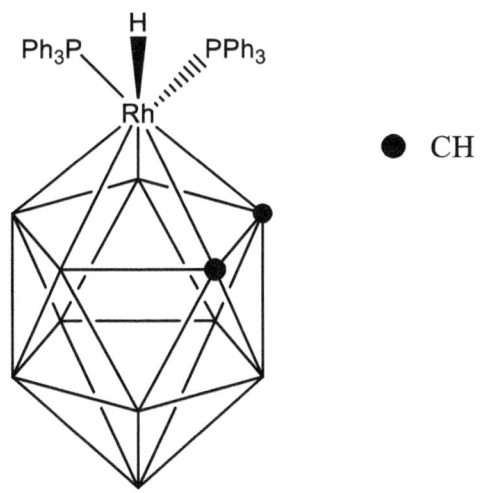

● CH

This concludes the chapter. By now you should be comfortable using Wade's Rules to determine the parent shape and the *closo-nido-arachno* state of a cluster. You should be able to work out the electron count of a transition metal fragment in the context of a cluster, though you might still need to check back to the equation for M occasionally.

Overall, you have hopefully made significant progress from where you started, and are becoming confident in the basic nuts and bolts of Wade's Rules problems even if you occasionally stumble.

The $M = V + L - 12$ procedure is a bit mysterious in the way I have chosen to present it. Yes, it works – but *why*? Appendix II deals with this question. Make sure you engage with your examiner's view of the topic to work out whether or not this is important to know deeply.

Interlude 3: ^{11}B NMR Spectroscopy

Like the carbon resonances in interlude 2, the NMR spectra of ^{11}B atoms in clusters can be a useful diagnostic tool. Again, the key piece of information in these spectra is the number of chemical environments.

The ^{11}B{^{1}H} NMR spectrum of a sample of B_5H_9 has two peaks. **Suggest a structure for this cluster.**

Wade's Rules predict a *nido* octahedron; the two environments (one apical BH group and four basal BH groups) are consistent with the two peaks in the spectrum.

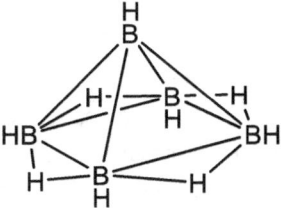

The ^{11}B{^{1}H} NMR spectrum of a sample of $Fe(CO)_3(C_2B_3H_7)$ has two peaks. **Suggest a structure for this cluster.**

This example was discussed in Chapter 3, where a 2,3-substitution pattern of carbon was presented in a *nido* trigonal bipyramid:

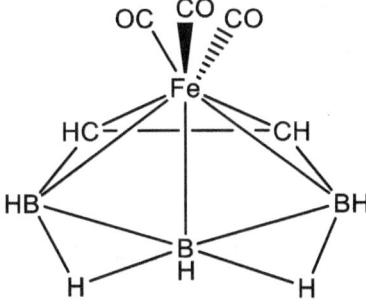

The B_3 fragment has one unique central boron and two flanking BH groups. These two chemical environments are consistent with the two environments seen in the spectrum.

Note that the 2,4-substitution pattern also results in two boron environments (if the bridging hydrides show significant fluxionality on the NMR timescale):

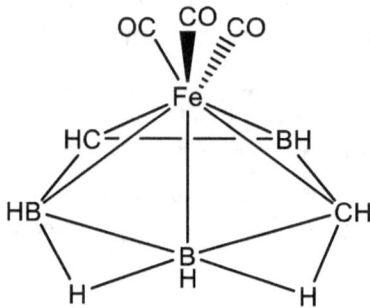

It is impossible to distinguish these isomers from the given NMR spectrum alone. In the absence of any reason to the contrary (e.g. a case study in the lecture course or additional data), most examiners would award marks for either answer so long as you justify it using the data.

The $^{11}B\{^1H\}$ NMR spectrum of a sample of $Sn(CR)_2B_4H_4$ has four peaks. **Suggest a structure for this cluster.**

There's quite a lot going on here.

Fragment	Skeletal electrons per fragment	Number of fragments	Total skeletal electrons
BH	2	4	8
CR	3	2	6
:Sn	2	1	2

The 8 pairs suggest a 7-corner shape (pentagonal bipyramid). As there are 7 terminal chemical fragments, this makes the structure *closo*.

The data requires all four boron groups to be distinct. Placing one in an apical site requires that the remaining three must be both in the ring and unique. The arrangement below accomplishes this because the BH groups at the 'back' of this representation are different: one is adjacent to Sn while the other is adjacent to basal CR.

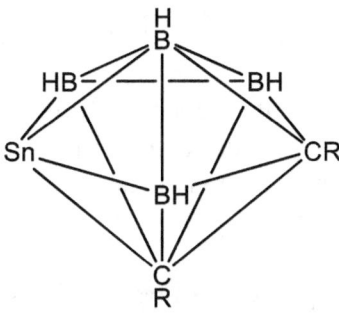

So the 2-Sn-4-CR substitution pattern above works, but note that this would also be true of the 2-Sn-3-CR isomer. Again, one technique isn't quite enough to properly pin the structure down.

In interlude 4 we will look carefully at a recent case study where multiple techniques are combined to characterise a carborane complex.

Chapter 4: Reactivity

This book focuses on structural aspects of clusters, and you should recognise this as a limit to the understanding you will develop using only this book. Purcell and Kotz is particularly good if you want to explore reactivity more deeply.

The problems in this chapter can all be solved by applying Wade's Rules to chemicals in reaction schemes. Be aware that there are (many) cluster reactions where this is not true.

The cluster compound $B_{10}H_{14}$ undergoes a two-electron reduction when reacted with KBH_4 in water. **Describe the structural changes which occur through the reaction.**

In this problem, the reaction is clearly framed as an invitation to examine the structures of the starting materials and products.

The starting material has BH fragments and bridging hydrogen atoms:

Fragment	Skeletal electrons per fragment	Number of fragments	Total skeletal electrons
BH	2	10	20
Bridging H	1	4	4

The twelve pairs of electrons suggest an eleven-corner shape, which will be *nido*. This parent shape resists easy description in words; in general, naming complex shapes is not needed to understand the Chemistry of large clusters.

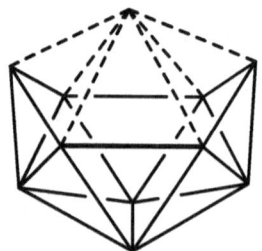

The simplest reading of the information in the question is that the product is $[B_{10}H_{14}]^{2-}$ (if the crude two-electron reduction has occurred). This cluster can be treated through the formal counting procedure (which I recommend), or you can note that the product has two more electrons than the starting material.

The thirteen pairs of electrons suggest a shape with twelve corners. This will be *arachno*. Drawing the bridging H atoms convincingly in perspective is hard.

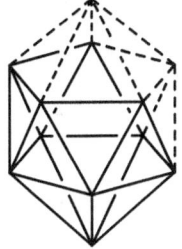 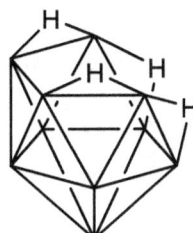

I like this example for two reasons.

1. It presents a clear demonstration of using structural models to solve reaction problems in cluster chemistry; and
2. It is a pure way of emphasising the 'opening up' of a cluster by adding electrons to the count. Adding a pair of electrons adds a corner to the parent shape.

Comment on the structural progression in the scheme below.

$$B_9H_9S \xrightarrow[\text{2 H}^+]{\text{2 e}^-} B_9H_{11}S \xrightarrow[\text{H}^+]{\text{2 e}^-} B_9H_{12}S^-$$

I normally dislike 'comment' as an instruction in an exam question because it is extremely vague. Here, though, it is directed more narrowly towards the structures of the clusters in the scheme. These structures can be worked out systematically.

Fragment	Skeletal electrons per fragment	Number of fragments	Total skeletal electrons
BH	2	9	18
:S	4	1	4

The eleven pairs of electrons suggest a 10-corner shape in the starting material (*closo* cluster).

Fragment	Skeletal electrons per fragment	Number of fragments	Total skeletal electrons
BH	2	9	18
Bridging H	1	2	2
:S	4	1	4

The twelve pairs of electrons suggest an 11-corner shape in the intermediate (*nido* cluster).

Fragment	Skeletal electrons per fragment	Number of fragments	Total skeletal electrons
BH	2	9	18
Bridging H	1	3	3
:S	4	1	4
Charge	1	1	1

The thirteen pairs of electrons suggest a 12-corner shape in the product (*arachno* cluster).

Addressing the 'comment' in the question might be done by sketching out the structures or writing a few sentences on the way that the B_9S skeleton is arranged over the corners of shapes with more and more corners as the cluster is reduced. Underlying all of this is the way that Wade's Rules distribute atoms according to parent shapes determined by the number of *electrons* within the cluster's skeleton.

The reactions above were reductions of the cluster skeleton, but there is also a huge wealth of acid/base cluster Chemistry.

Explain the regioselectivity of the transformation below.

$$1,2\text{-}C_2B_{10}H_{12} \xrightarrow[\text{2. RX}]{\text{1. BuLi}} 1,2\text{-}(CR)_2B_{10}H_{10} + 2LiX$$

The *closo* 12-corner cluster is substituted at the carbon atoms rather than the boron atoms. **Broadly, you must decide whether you think the explanation for this is thermodynamic or kinetic.**

A kinetic explanation will focus on the easiest processes at each step. This type of explanation is the dominant model in organic chemistry ('mechanisms').

A thermodynamic explanation will focus on how the product is the most stable arrangement of atoms given all the possibilities. This type of explanation is dominant in Inorganic Chemistry (probably because the bonds are normally weaker).

If you commit to a thermodynamic argument, then you should **find a way that the observation** (an exo C–C bond in the product) **should be preferred over the possible alternatives** (an exo B–C bond in the product).

[Note: formally, comparing all products with all starting materials might also be productive in certain problems; here the lattice enthalpy of LiX would be a significant feature of any such discussion.]

An argument for this might run something like:

"The transformation selectively substitutes at the carbon atoms because of bond strength. A (C–C)σ bond is stronger than the alternative (B–C)σ bond because the perfect energy match of the carbon atomic orbitals leads to a larger perturbation."

[Note how this contrasts with the argument made in Interlude 1, where *cis*-C_2 configurations *within* the cluster skeleton were disfavoured because of the repulsion between atoms bearing a partial negative charge.]

I have done a couple of things in my answer which are worth focusing on for exam technique:

1. Mirrored language in the question ('transformation', 'selective') to emphasise how I have addressed the problem; and
2. Committed clearly to a bond strength argument (rather than tried to 'hedge my bets').

You might have advanced a kinetic argument instead. To do this, you would have to identify a specific step which sends the reaction in a particular direction.

For example, you might focus on the deprotonation step:

"The deprotonation is selective for C–H bonds because they are more acidic than B–H bonds. The energetic barrier to the E–H deprotonation step is therefore lower when E is C, leading to a faster reaction."

I personally favour the kinetic argument: it aligns better with my intuitions about the behaviours of hydrides near the top of the p-block. Both arguments are logically self-consistent, though.

The cluster compound B_5H_9 (A) reacts with one equivalent of KH to form an anionic species (B). **Suggest structures for (A) and (B).**

Wade's Rules can be applied to the starting material:

Fragment	Skeletal electrons per fragment	Number of fragments	Total skeletal electrons
BH	2	5	10
Bridging H	1	4	4

This suggests (A) is a *nido* octahedron:

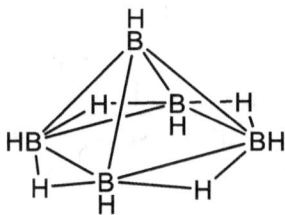

KH acts as a base, and there are three possible sites for the deprotonation. **How might you decide which type of proton is lost?**

Again, applying Wade's Rules to $[B_5H_8]^-$ will tell you which product is the thermodynamic one.

Fragment	Skeletal electrons per fragment	Number of fragments	Total skeletal electrons
BH	2	5	10
Bridging H	1	3	3
Charge	1	1	1

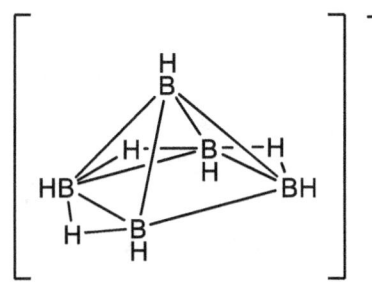

In general, bridging hydrogen groups are more acidic than the terminal BH groups. The bonding scheme behind this selective deprotonation is discussed at the end of Appendix I if you wish to explore this idea in more depth.

$[Sn_9]^{4-}$ reacts with one equivalent of $(1,3,5\text{-}Me_3C_6H_3)Cr(CO)_3$ to give a tetraanionic product with no signals in the ^1H NMR spectrum and three fundamental (C–O) stretches in its IR spectrum. **Suggest a product for this reaction.**

There is a lot of data in the question here. Breaking this down systematically is an important step in solving this problem. **What does the absence of ^{1}H NMR signals indicate?**

The only protons in the starting substances were all in the arene fragment of the chromium complex; the arene must be absent from the product.

What does the presence of three (C–O) stretches indicate?

The starting material contained three carbonyl groups. While symmetry considerations may reduce the number of observed IR peaks (the starting material shows two stretches transforming as $A_1 + E$ in the C_{3v} point group), no amount of symmetry can increase the number of peaks. Strictly, the IR data therefore suggests there are *at least* three carbonyl groups in the product. When compared with the 'one equivalent' information in the question, this suggests that there are exactly three.

Identify the fragments and suggest a structure for the product.

Assuming that a $Cr(CO)_3$ fragment is preserved in the product

$$M = V + L - 12$$

becomes

$$M = 6 + 6 - 12 = 0$$

Fragment	Skeletal electrons per fragment	Number of fragments	Total skeletal electrons
:Sn	2	9	18
$Cr(CO)_3$	0	1	0
'tetraanionic' charge	1	4	4

The eleven pairs of electrons suggest a 10-corner shape, which is populated with 10 chemical fragments (*closo*):

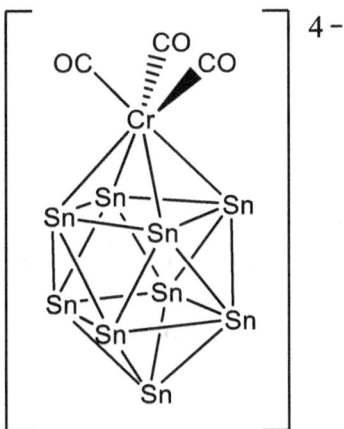

Note that the *nido* Zintl cluster has *acted as a nucleophile* towards the chromium centre; arguably it has *chelated* the transition metal.

This concludes the chapter. Most of these reactivity problems were structural problems in disguise, and I have tried to help you see how to use Wade's Rules to address common terminology used in exam questions.

In my practical experience of inorganic reactions, weird stuff happens all the time; it's normally more important to be able to analyse after the fact than predict anything before it.

The mechanisms which form such a large part of organic courses rely upon kinetic features (e.g. breaking the strong C–C σ-bond) dominating reaction outcomes. Inorganic systems often rely more heavily upon weaker bonds, and have more opportunities to discover their thermodynamic minima.

This thermodynamic 'sink' means that many examinable reaction problems in this topic involve applying Wade's Rules to both the starting materials and the products. There are other reactions of clusters which do not follow this pattern; you should see whether your lecturer includes them in their teaching (including reading lists) or assessment.

Interlude 4: Advanced Characterisation

Note: this material is very advanced, and will challenge you. I intend it as an illustration rather than as core content. Don't be scared!

The previous two interludes used only one type of NMR spectroscopy, but in rigorous research chemistry you would normally show how a range of techniques build up a compelling argument for your molecule having a certain structure (this is sometimes described as a 'mosaic' approach to characterisation).

This interlude presents examples from a recent research article to help you see what professional characterisation looks like.

I chose Riley et al *Dalton Trans.*, 2016, **45**, 1127. This paper reports several things, but I want to focus on the use of a deprotonated bis(carborane) as a chelating ligand. The reaction scheme is presented overleaf, with BH units left unlabelled in the deltahedra for clarity. I have taken the characterisation data for Compound A from Yang et al in *Inorg. Chim. Acta*, 1995, **240**, 371-378.

Both of these papers have been reviewed by peers (other academics in closely-related areas). They are high-quality pieces of research. I have neglected the mass specrometry and elemental analysis data as I wanted to focus on the NMR assignment.

It is a tiny anecdote, but I have personally made the ruthenium starting material in this reaction. The arene (*p*-cymene) is also present in the basil plant, and the air-stable metal complex smells nice: floral, but not sweet. Presumably this is because the substance decomposes, releasing the uncomplexed cymene in trace amounts.

$[RuCl_2(cymene)]_2$ is also brown, which is a curious phenomenon in synthetic chemistry. No pure pigment is brown (you may remember making brown paint from *mixing* all the other colours when you were an annoying child), which normally suggests that brown products are somehow impure. I suspect that as a powder it exists in some sort of poorly-defined polymeric form; in solution it seems to react fine.

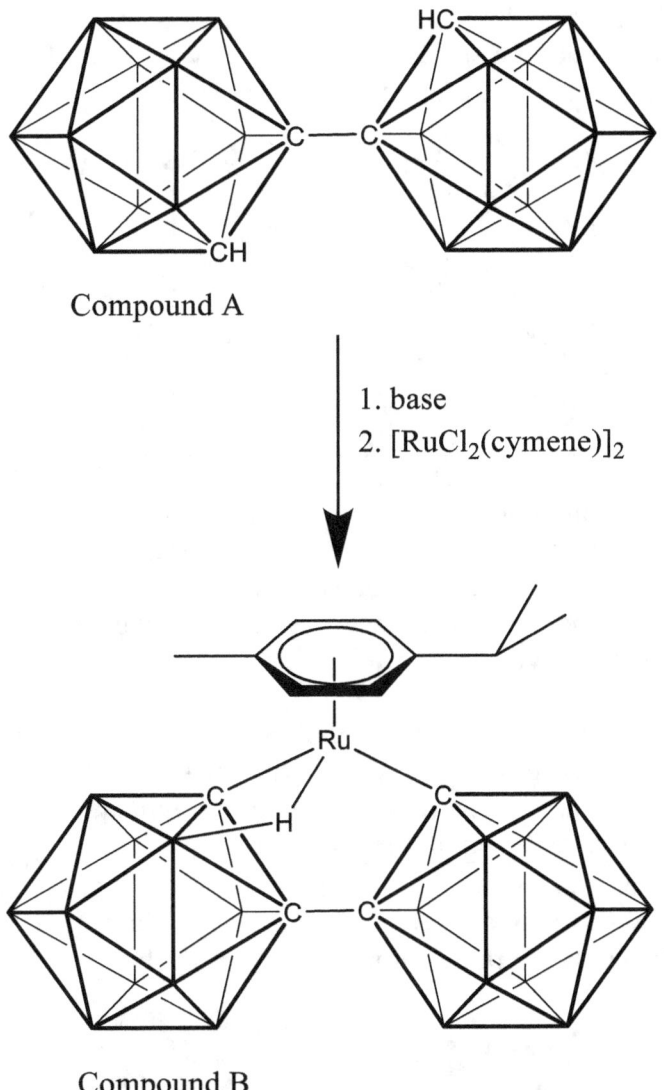

Compound A

1. base
2. [RuCl$_2$(cymene)]$_2$

Compound B

The ^{11}B{^{1}H} NMR spectrum of Compound A (molecular formula [C$_2$B$_{10}$H$_{11}$]$_2$) shows six environments. The ^{13}C{^{1}H} spectrum shows two environments. **Is this consistent with the structure presented?**

I think it is. Remembering that the C–H group is distinct from the C–C group bridging the deltahedra (giving two carbon environments), it is possible to map out the six BH environments with a bit of thought about the mirror plane through the molecule:

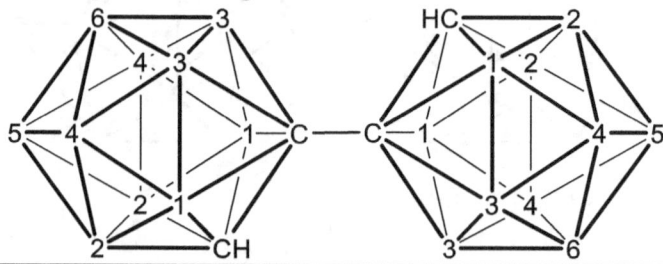

The ^{1}H NMR spectrum of Compound A shows only one signal. **Is this consistent with the structure presented?**

Yes, but this might be unclear at first. The CH group is resolved, but the BH protons are not. It is quite common for protons bonded to quadrupolar nuclei (such as both ^{10}B and ^{11}B) to disappear. Briefly, this is because the spin of the ^{1}H nucleus can relax by coupling to the adjacent quadrupolar nucleus rather than by emitting an electromagnetic wave.

This actually makes the observation of one proton signal quite compelling evidence when compared with the six boron signals and the two carbon signals. It is hard to see how this combination of evidence could be constructed without the proposed structure.

Finally, the crystal structure of Compound A is shown below. **Is this consistent with the proposed structure?** (note that C atoms are shown in black, B in grey; H atoms were not resolved)

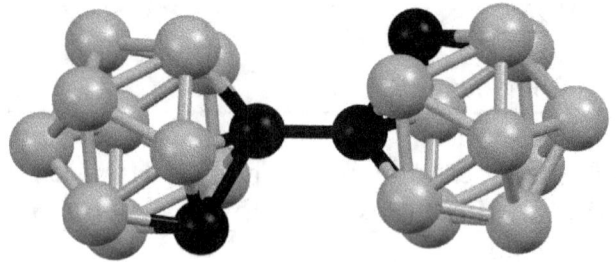

Crystal structures are an interesting artefact. On the one hand, yes: the presented data matches the proposed structure (though the perspectives

54

can be hard to 'get' when a 3D structure is shown on a 2D page). It can also be experimentally difficult to tell atoms with similar masses apart.

But crystal structures should always be treated with a bit of caution. They represent the solid-phase structure (which may not match the solution-phase behaviour). They also report the reaction components which crystallised, which may not be the dominant reaction outcome. In this particular case, the H atoms were not resolved, so we cannot infer anything about the arrangement of H atoms from the structure alone.

In the context of the other data, though, this structure of the bis(carborane) skeleton makes the characterisation even more compelling. Each piece of data pushes the weight of evidence a little bit further, and the overall argument is strong.

Compound A is deprotonated with a base and then reacted with the Lewis acidic metal centre to give Compound B.

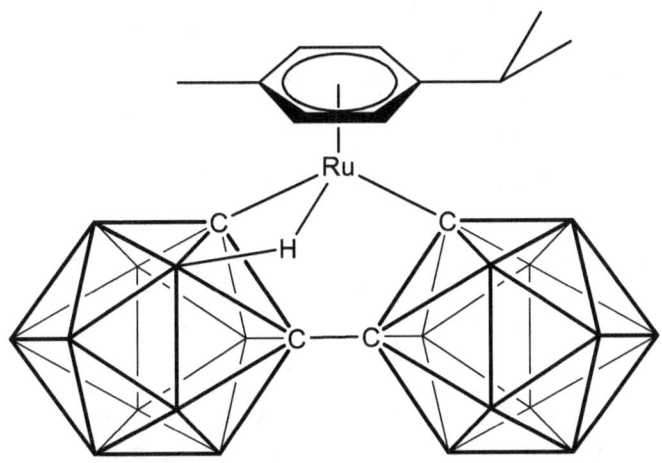

The $^{11}B\{^1H\}$ NMR spectrum of Compound B shows five peaks, one of which has a shoulder. **Is this consistent with the proposed structure?**

At first sight, no. The agostic hydrogen renders the deltahedra inequivalent in the static structure above: there should be *many* more boron environments than the six seen in the starting material. **How might this observation be reconciled with the characterisation?**

The argument the authors propose is fluxionality. They suggest that the agostic hydrogen might come from one of four BH groups (labelled 1 below). The metal switches between each group rapidly, so the NMR spectrum 'blurs' those four groups into one environment. Fluxionality is discussed in a molecular context in *Assigning Inorganic NMR Spectra* by O'Neill.

As a way of reconciling the data, fluxionality seems plausible (though it would have to match the other data as well). This mechanism would, by my reading, give six distinct boron environments.

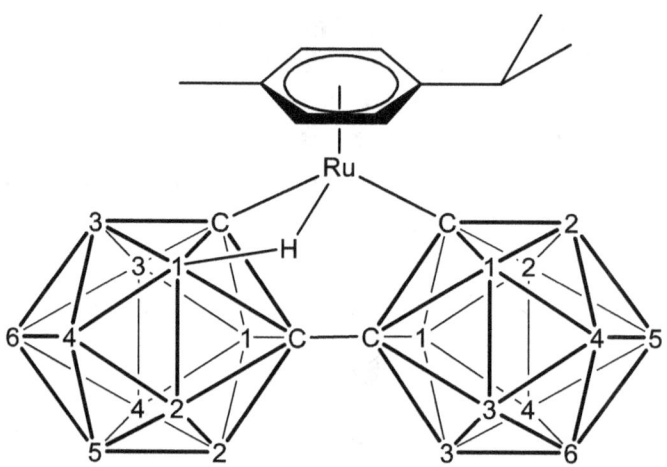

When compared with the 5-plus-shoulder spectrum observed, predicting six peaks seems ok if the shoulder is actually two distinct-but-close peaks. By itself, the $^{11}B\{^1H\}$ data isn't enough to characterise the molecule. It's consistent with the proposed structure, though.

The ^1H NMR spectrum can be compared with the $^1H\{^{11}B\}$ NMR spectrum. **What types of signal will appear once boron decoupling is applied?**

Protons attached to boron atoms (specifically ^{11}B atoms) will produce signals in the $^1H\{^{11}B\}$ NMR spectrum. Comparing the two spectra therefore gives a clear indication of which signals derive from BH groups.

Six additional peaks appear once the $\{^{11}B\}$ decoupling pulse is applied. **Is this consistent with the proposed structure?**

Yes. The six boron environments proposed above are supported by the appearance of six distinct H signals coupled to ^{11}B atoms. Again, the mosaic of evidence supports a case of support for the proposed structure.

The $^1H\{^{11}B\}$ experiment was run at two different temperatures: 298 K and 203 K. The 4H peak at –0.02 ppm assigned to the agostic BH group in the room temperature spectrum splits into two signals at low temperature: 0.78 ppm (2H) and –1.03 (2H). **Is this consistent with the proposed fluxionality between four BH groups?**

Cooling the solution lowers the population of molecules capable of reaching the activation energy for this intramolecular process. This should 'freeze' the fluxionality. In the perfectly static structure, all the BH groups labelled below should be distinct.

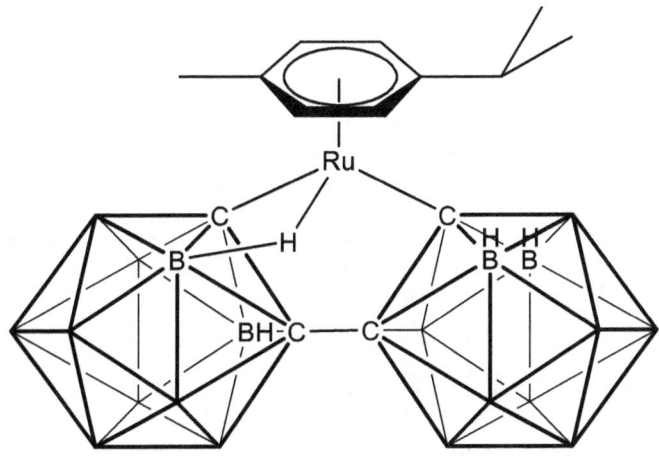

This should give 4 peaks, but only 2 are observed at 203 K. The authors suggest that the fluxional process is only partially arrested, which seems plausible. By detailed comparison with the arene signals, they present a robust argument for the still-permitted fluxionality being between BH groups on the same cage.

The crystal structure of Compound B is presented below. **Does this support the proposed structure?**

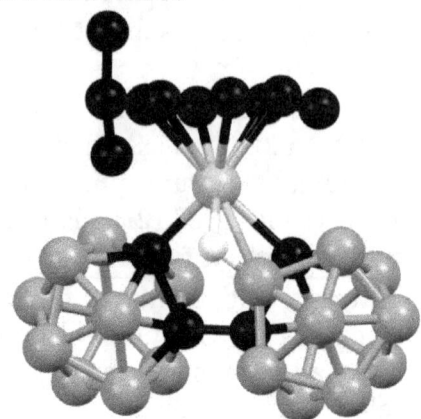

(C atoms are shown in black; B in grey; Ru in [slightly] lighter grey; H atoms were all resolved, but only the agostic one is shown in white for clarity)

Again, the crystal structure must be interpreted carefully. The fluxionality proposed earlier to explain the number of boron peaks is compatible with this structure as a 'snapshot'.

Compound B seems not to have been characterised using ^{13}C NMR spectroscopy. There is other data which I haven't presented here (e.g. mass spectrometry), but the absence is worth a brief discussion. **What additional information would the ^{13}C{^{1}H} NMR spectrum reveal?**

Perhaps you have a different opinion, but I think the carbon NMR spectrum would have added a little bit of weight to the mosaic approach of characterisation. That said, the overall case seems strong enough to me at this stage, and the information would likely be redundant after the ^{1}H NMR spectrum has been compared with the ^{1}H{^{11}B} spectrum.

This was a brief tour of one reaction from the recent literature. Its main purpose was to show you how a 'number of peaks' analysis can be useful in real research, and to give you a taste of the complexity of forming a 'mosaic' judgement about standards of characterisation. This way of thinking about synthetic chemistry takes a long time to get used to, and you will likely have found the detailed analysis very challenging. It may even have been quite uncomfortable; this is normal. The next chapter focuses on more examinable material.

Chapter 5: Advanced Problems

What makes a Wade's Rules problem difficult?

You may have an insight into the sorts of mistakes you keep making, and I hope you take some time to think about how you might address your particular grasp of this topic – your answer to this question is extremely important.

In my experience of teaching Wade's Rules, there are a few characteristic features of problems which students find difficult:

1. The cluster has a lot of different fragments in it (perhaps three or more, maybe also with a charge);
2. The fragments are a bit weird (transition metals, bridging hydrides, exo lone pairs);
3. The clusters are big;
4. The problem challenges something about your knowledge (e.g. presents you with data contradicting the 'perfect' Wade's Rules structure).

1, 2, and 3 are problems which derive from information processing. In Chemistry Education, this is commonly discussed using a model called Cognitive Load. Cognitive Load's central idea is that you can only process so much information at a time. The way to beat this problem is practice; this helps you deal with problems as one process rather than lots of bitty details (a mental manipulation sometimes called 'Chunking' the material).

4 is slightly different, but a much better reflection of what Chemists do every day. To score highly in these problems you need academic confidence in your Wade's Rules analysis, the insight to know that the thermodynamic structure is not always the experimental structure, and a good grasp of how the spectroscopic techniques link to features in the structures. It is normally the spectroscopy bit that students struggle with most; I believe that this is because spectroscopy is normally taught in a different course (by a different person). This book has tried to deliberately weave in spectroscopy through the interludes, developing your application of NMR to cluster characterisation.

There are other ways a non-problem exam question can be challenging, of course. Some questions are difficult to do in time. Some questions test raw recall, and you might not be able to remember a key fact about (e.g.) a molecular orbital scheme. I find these questions boring, myself. I don't think they test valuable learning.

The problems in this chapter are good challenges for students after a Wade's Rules course. They are supposed to be hard, but I hope you find some of them satisfying as well.

Suggest structures for the three molecules listed below.

$$Ru_3(CO)_{12} \qquad Ir_4(CO)_{12} \qquad Rh_6(CO)_{16}$$

$Ru_3(CO)_{12}$ can be broken neatly into three $Ru(CO)_4$ fragments. So

$$M = V + L - 12$$

becomes

$$M = 8 + 8 - 12 = 4$$

Fragment	Skeletal electrons per fragment	Number of fragments	Total skeletal electrons
$Ru(CO)_4$	4	3	12

An *arachno* trigonal bipyramid presents as a triangle of $Ru(CO)_4$ units.

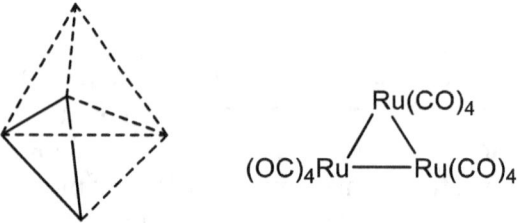

The iridium species can be considered as four $Ir(CO)_3$ fragments;

$$M = V + L - 12$$

becomes

$$M = 9 + 6 - 12 = 3$$

Fragment	Skeletal electrons per fragment	Number of fragments	Total skeletal electrons
Ir(CO)₃	3	4	12

The skeleton is again an *nido* trigonal bipyramid, but with Ir(CO)₃ groups on each corner. The analogy between Ru(CO)₄ and Ir(CO)₃ is formalised in Appendix III.

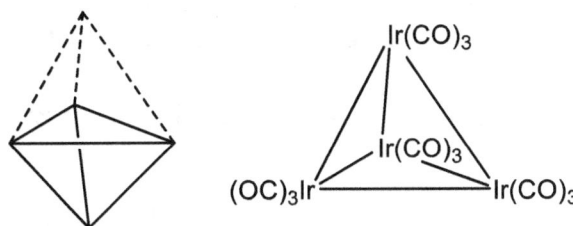

Rh₆(CO)₁₆ is harder to split into fragments because the ratio of metal to ligand sits between two integers (2 and 3).

The Rh(CO)₂ and Rh(CO)₃ fragments can be considered separately.

For the two dicarbonyl fragments
$$M = V + L - 12$$
becomes
$$M = 9 + 4 - 12 = 1$$

For the four tricarbonyl fragments
$$M = V + L - 12$$
becomes
$$M = 9 + 6 - 12 = 3$$

Fragment	Skeletal electrons per fragment	Number of fragments	Total skeletal electrons
Rh(CO)₃	3	4	12
Rh(CO)₂	1	2	2

The structure therefore adopts a *closo* octahedral skeleton.

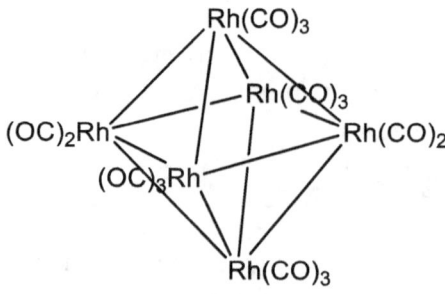

The synthesis of carborane clusters is commonly accomplished by a two-step procedure involving Lewis bases such as Et_2S. **Suggest structures for all cluster compounds in the scheme below.**

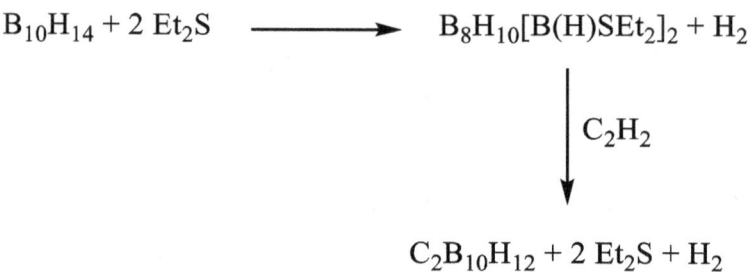

$$B_{10}H_{14} + 2\ Et_2S \longrightarrow B_8H_{10}[B(H)SEt_2]_2 + H_2$$

$$\downarrow C_2H_2$$

$$C_2B_{10}H_{12} + 2\ Et_2S + H_2$$

The structures of $B_{10}H_{14}$ (*nido* 11-corner shape) and $C_2B_{10}H_{12}$ (*closo* 12-corner shape) should be quite accomplishable for you by now. Experimentally, the final carborane product has a 1,2-arrangement of carbon atoms (kinetic product) by this low-temperature route.

$B_8H_{10}[B(H)SEt_2]_2$ is a more complex species because of the $[B(H)SEt_2]$ fragment. **How might you start to think about the number of skeletal electrons provided by $[B(H)SEt_2]$?**

Hint: When considering the 'naked' S atom, we allocated a lone pair exo to the cluster and counted the leftover electrons (four).

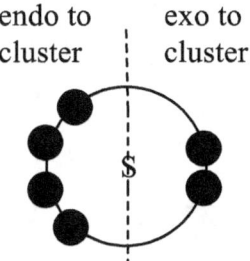

endo to cluster   exo to cluster

This contrasted with BH, which left two electrons for skeletal bonding after accounting for the exo Lewis bond:

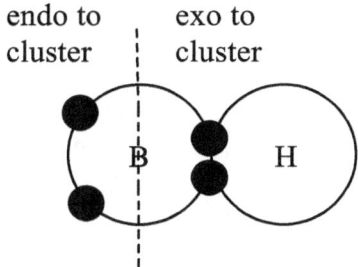

endo to cluster   exo to cluster

How might this approach be extended to [B(H)SEt₂]?

Sketching the immediate coordination sphere around S (and ignoring C–C and C–H bonding) shows that there are still four electrons donated into the cluster skeleton.

You might have suggested an analysis which donates two electrons to the skeleton:

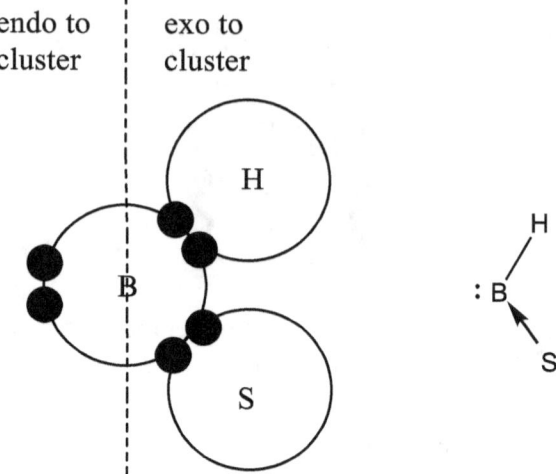

In a really unusual twist, though, the B–H bond actually sits *within* the cluster skeleton. This means that its electrons contribute to the skeletal count. Note that *there is no way you could have known this before being told.*

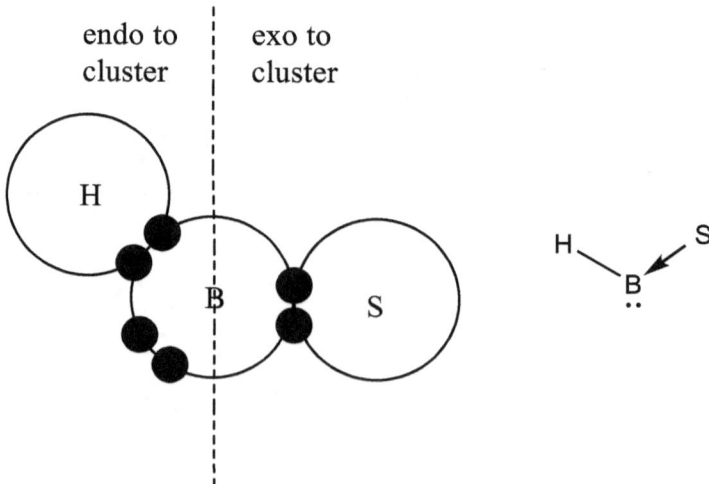

The fragment therefore donates 4 electrons into the cluster skeleton.

Fragment	Skeletal electrons per fragment	Number of fragments	Total skeletal electrons
BH	2	8	16
Bridging H	1	2	2
B(H)SEt$_2$	4	2	8

The thirteen pairs of electrons suggest a shape with 12 corners. There are only 10 terminal chemical fragments, so this will be *arachno*. Note how the B–H bond of the [B(H)SEt$_2$] fragment sits: not pointing away from the cluster radially, but lying along the surface of the skeleton. This structural feature allows the electrons in the B–H bond to contribute to the skeletal electron count.

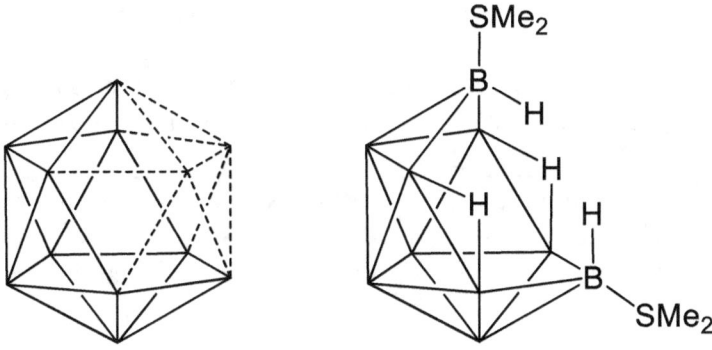

The *arachno* 'openness' of this structure is actually why the second step of this reaction scheme works; ethyne reacts with the exposed, electron-rich open face (it doesn't react with the less-open *nido* face of the starting material) at temperatures low enough to preserve the C–C bond through the process.

Clusters **A** and **B** can be made according to the scheme below.

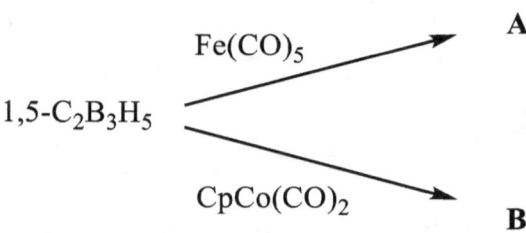

A shows two peaks in its $^{11}B\{^1H\}$ NMR spectrum and two peaks in its $^{13}C\{^1H\}$ NMR spectrum. Its IR spectrum contains three stretches attributable to carbonyl groups.

B shows two peaks in its $^{11}B\{^1H\}$ NMR spectrum and two peaks in its $^{13}C\{^1H\}$ NMR spectrum, one of which is around the aromatic region. Its IR spectrum does not contain any carbonyl stretches.

Propose reasonable structures for compounds A and B consistent with the reported spectroscopic data. Comment on your answer in light of the isolobal principle.

Reasonable is exam code for *using Wade's Rules*, so the first step is to identify the fragments. This is probably most straightforward for the cobalt system, which has no carbonyls and some aromatic carbon.

Fragment	Skeletal electrons per fragment	Number of fragments	Total skeletal electrons
BH	2	3	6
CH	3	2	6
CoCp	2	1	2

The seven pairs of electrons suggest a six-corner shape (*closo* octahedron. Preserving the separation of carbon atoms in the 1,5-starting material gives a cluster with one carbon and two boron environments, so that's what I've proposed.

The iron system is a little harder to address. The three carbonyl IR signals suggests the presence of $Fe(CO)_3$. Knowing this, we can perform the electron count.

Fragment	Skeletal electrons per fragment	Number of fragments	Total skeletal electrons
BH	2	3	6
CH	3	2	6
$Fe(CO)_3$	2	1	2

The carbon NMR spectrum reports one cluster and one carbonyl environment. Again, I have suggested that the metal centre adds so as to preserve the separation of carbon atoms.

The isolobal comment is tricky. 'Comment' is a vague exam instruction, and you should pay careful attention to cues such as the marks available when working out how much to write.

Broadly, the isolobal principle speaks to the way that inclusion of the CoCp and $Fe(CO)_3$ fragments results in isostructural clusters (i.e. both **A** and **B** are *closo* octahedral). To do this, they must provide two things:

1. The same number of electrons to the cluster (2 electrons); and
2. Frontier orbitals for bonding in the cluster skeleton with:
   a. Similar energy (though not necessarily in the same order of energy levels); and
   b. The same symmetry.

A discussion of isolobality is developed in Appendix III; it hasn't been a feature of the core content in this book, but it is satisfying to think about.

Stepping back from the question a little bit, it is nice to note that BH is isolobal with both metal centres, too.

The cluster $(Et_3NB)PB_8H_8$ displays six signals in its $^{11}B\{^1H\}$ NMR spectrum. **Suggest a structure for the cluster which agrees with Wade's Rules and the NMR data.**

Similar to the SEt_2 example earlier, we have to think carefully about the Et_3NB fragment. **Work out how the electrons are arranged around the boron atom.**

The dative bond from nitrogen means that the boron atom can provide three electrons for bonding in the cluster skeleton:

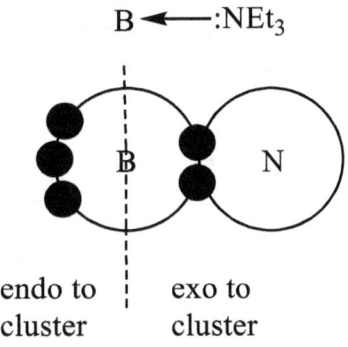

$$B \longleftarrow :NEt_3$$

endo to cluster $\qquad$ exo to cluster

Knowing this, **perform the electron count.**

Fragment	Skeletal electrons per fragment	Number of fragments	Total skeletal electrons
BH	2	8	16
:P	3	1	3
BNEt$_3$	3	1	3

The eleven pairs of electrons suggest that the structure will have ten corners. The ten terminal fragments mean that this will be *closo*.

Arrange the fragments to agree with the NMR data.

I have numbered the BH groups into NMR environments for clarity, intending that BNEt$_3$ would be environment 1. The equivalence of the 4 groups and 5 groups can be hard to see at first, but there is a mirror plane within the molecule.

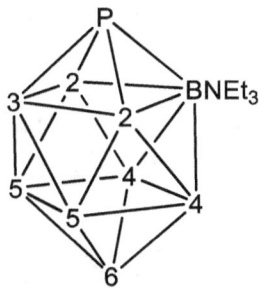

Predict the shape of $B_3H_7[Fe(CO)_3]_2$.

The fragments are varied, but not too hard to identify.

Fragment	Skeletal electrons per fragment	Number of fragments	Total skeletal electrons
BH	2	3	6
Bridging H	1	4	4
[Fe(CO)₃]₂	2	2	4

Seven pairs of electrons suggest a parent shape with 6 corners. Experimentally, the resulting *nido* octahedron has iron atoms in both the apical and basal positions (the *cis* isomer).

Predict the structure of $[B_6H_6]^{2-}$.

Hopefully you predicted a *closo* octahedron.

This was the first structure in the book, way back in Chapter 1.

I hope that including it as the last one in the programme would help you to reflect upon how far you've come. It's so easy for students to feel that they're not making progress because learning forces you to steer into your ignorance. Looking back from time to time can help you stay motivated.

And that's it! You've finished the book; I hope you found it useful. If you are looking to read further, I feel that the best book on this topic is *Boranes and Metallaboranes* by Housecroft. It is just wonderful.

The big general textbooks have good treatments, and I can particularly recommend Greenwood & Earnshaw for structure and Purcell & Kotz for reactivity. In terms of specialist texts, I have a soft spot for the out-of-print *Introduction to Cluster Chemistry* by Mingos and Wales, which is very dense but also very thorough. Housecroft's *Cluster Molecules of the p-Block Elements* is good if you want to consider the electron-deficient clusters described by Wade's Rules in the context of clusters which are electron-precise or electron-rich.

In terms of preparing for assessments, I suggest you attend carefully to what your instructors emphasise within the topic. There are *so* many different ways of designing a course which includes Wade's rules, and so many ways of assessing them. I have focused heavily on structural features of clusters and used the interludes to compare structures with spectra, but have not dug deeply into the reactivity of these systems. Your instructor might take a different view – make sure you know where they're coming from.

Resources like past paper questions are also valuable in helping you understand what aspects of the topic are valued by your examiners. I strongly suggest that part of your preparation involves doing past papers *timed* and *without your notes*. This will be hard, but will force you to prepare for the weirdly specific task of doing an exam rather than worthy but ill-defined goals like being a good scientist.

The literature on clusters is still developing. If you find yourself looking for chances to practice 'real life' assignment, your University will have subscriptions to some relevant journals with recent publications. Reading papers is extremely disorienting at first (no-one really talks about that?), but there will normally be a 'synthesis and characterisation' section which you can focus on. I suggest 'carborane' and 'zintl' as useful search terms for narrowing the field of papers.

Good luck! Please write to me if you found this book helpful, or if you want to suggest any improvements.

MON, Oxford, April 2020

michaeloneill.org

Appendix I: The Molecular Orbital Basis of Wade's Rules

The arrangement of orbitals in $B_6H_6^{2-}$ has been studied intensively. Considering the BH fragment as being *sp*-hybridised results in three bonding combinations of orbitals. For comparison, these are shown below alongside the corresponding atomic wavefunction of the same symmetry in an octahedron (no antibonding orbitals are shown).

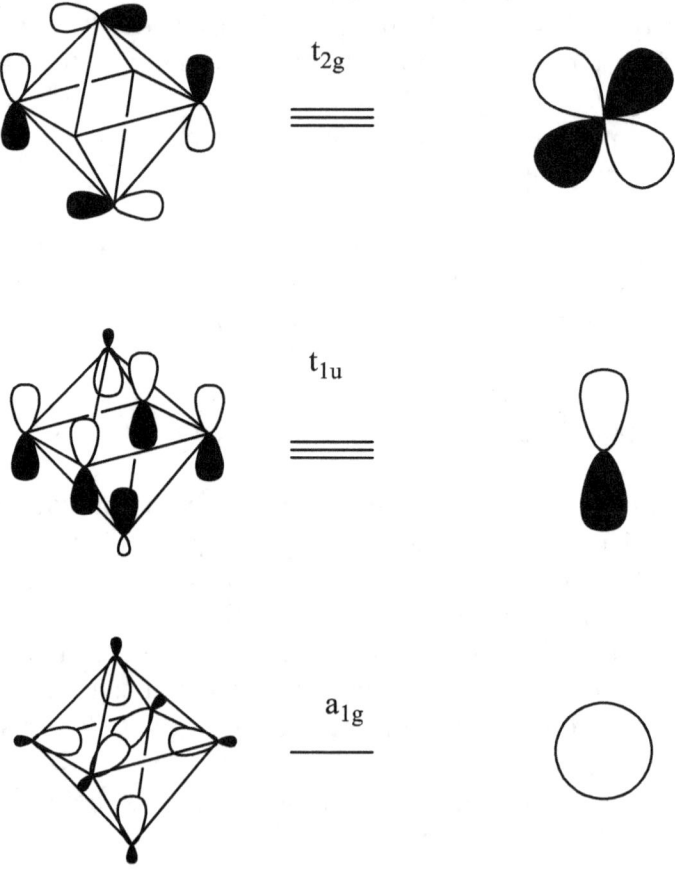

The seven electron pairs of the *closo* octahedron fill all of these bonding molecular orbitals. This argument is similar to the ones presented for Hueckel aromaticity or the eighteen electron rule: *filling all the bonding orbitals leads to a stable molecule.*

This Appendix will consider what happens to each of these orbitals as the *closo* octahedron (in the O_h point group) loses corners to become *nido* (C_{4v}) and *arachno* (C_{2v}) octahedra.

First, the most symmetric MO is conserved quite cleanly through the descent in symmetry.

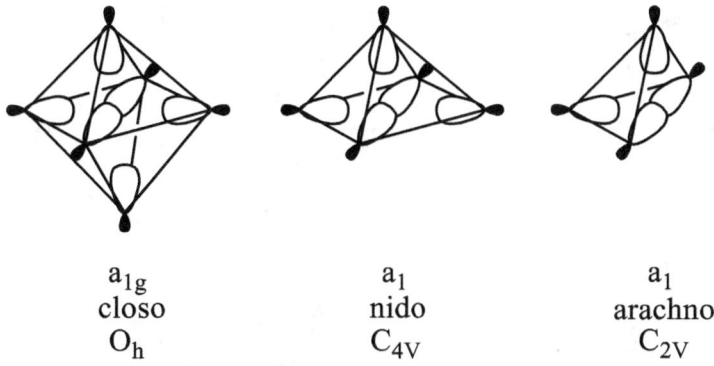

$$a_{1g}$$
closo
$$O_h$$

$$a_1$$
nido
$$C_{4V}$$

$$a_1$$
arachno
$$C_{2V}$$

The t_{1u} orbitals are affected more significantly, and the triply degenerate set splits into symmetrically different environments as objects along the z and y directions are removed. Note that degenerate molecular orbitals are grouped inside square brackets.

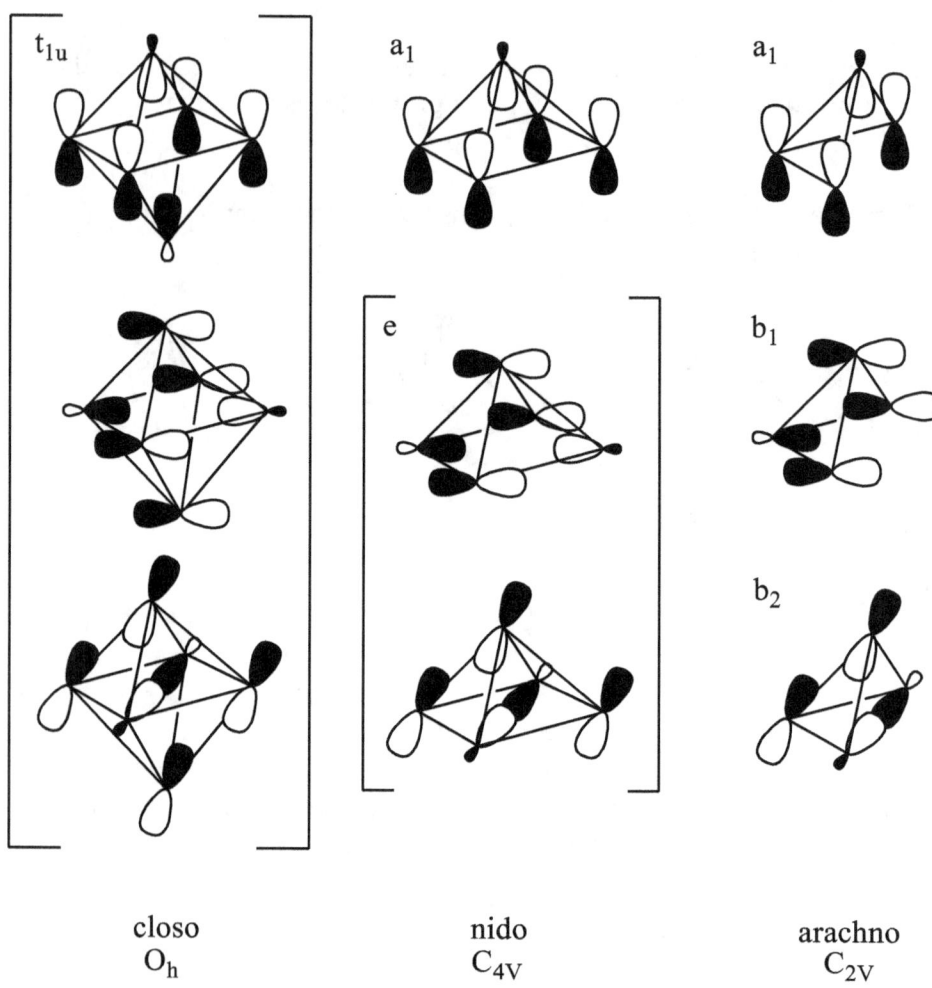

closo	nido	arachno
O_h	C_{4V}	C_{2V}

The basis for how Wade's Rules work is that there were three bonding MOs in the *closo* structure (as a triply degenerate set) and there are *still* three bonding MOs in the *nido* cluster (as one singly and one doubly degenerate sets) and there are *still* three bonding MOs in the *arachno* cluster (as three singly degenerate sets). In all three cases, these bonding MOs can accommodate six electrons despite the loss of atoms.

The descent of the t_{2g} set is harder to see spatially, but follows a similar pattern using only p-orbitals tangential to the cluster radius:

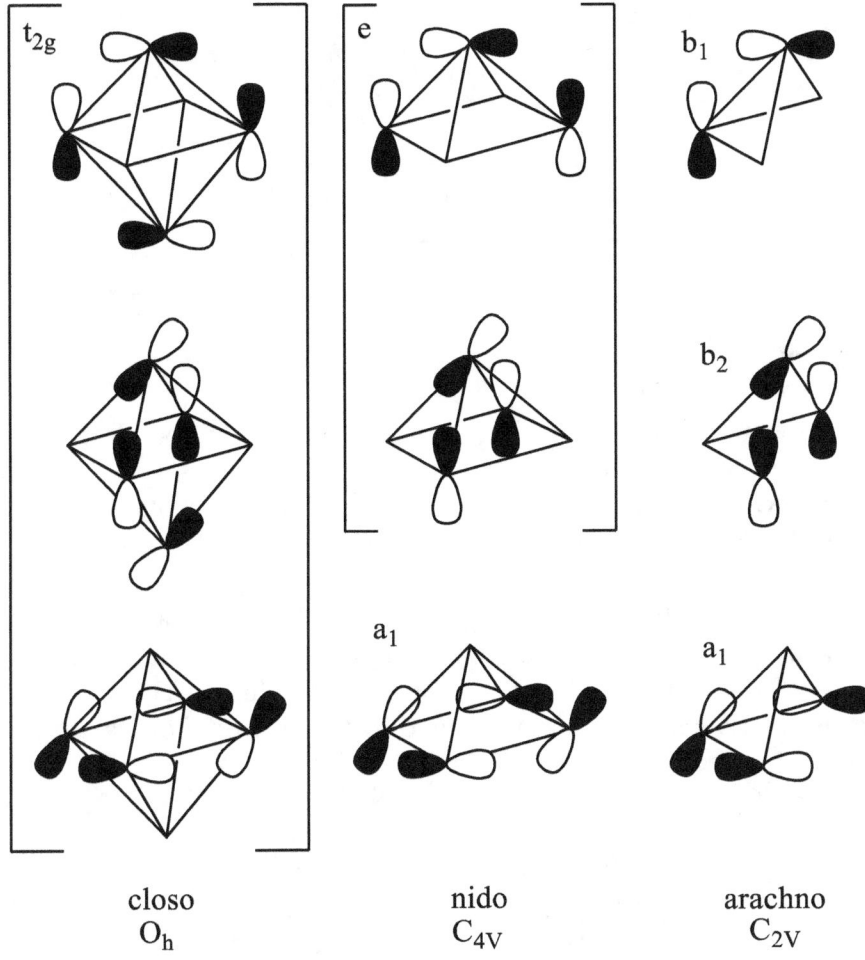

| closo | nido | arachno |
| O_h | C_{4V} | C_{2V} |

Again, the core point is that these molecular orbitals remain *bonding* orbitals even through the process of losing corners. In each of the *closo*, *nido*, and *arachno* cases we can formally count the number of bonding electrons which can be held in these orbitals.

Closo		Nido		Arachno	
a_{1g}	2	a_1	2	a_1	2
t_{1u}	6	a_1	2	a_1	2
		e	4	b_1	2
				b_2	2
t_{2g}	6	a_1	2	a_1	2
		e	4	b_2	2
				b_1	2
Total	**14**		**14**		**14**

In a cluster with this sort of MO scheme, seven pairs of bonding electrons is therefore consistent with a six-corner parent shape. This is true even when there are fewer than six chemical fragments to decorate the corners, leading to *nido* and *arachno* structures.

Note that the descent in symmetry of the t_{1u} set provides a neat explanation for the observation that bridging hydrogen atoms are typically located around the open face of a *nido* cluster. The a_1 orbital (left) is clearly stabilised when bridged by the 1s orbital of a hydrogen atom (right).

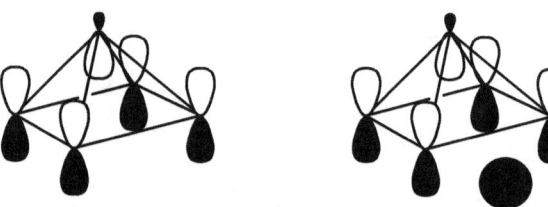

This is seen in the location of hydrogen atoms in structures such as B_5H_9.

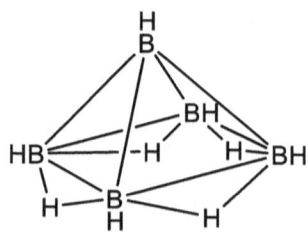

Appendix II: The Transition Metal Fragment Electron Count

Broadly, the electron count M can be described as

metal electrons + ligand electrons − uninvolved electrons

Specifically, the formula used to calculate how many electrons a metal fragment contributes to a cluster is

$$M = V + L - 12$$

where V is the valence count of the metal (the group number), L is the number of electrons donated by the ligand using the neutral organometallic counting scheme, and 12 is a constant.

The way that a metal's group adds electrons to the count seems reasonable, as does the contribution of ligand electrons to the count.

The 12 is a real puzzle, though: why are 12 electrons subtracted from the count in a way that is independent of the actual structure? Surely the number of electrons 'taken up' by M–L bonds should relate to the number of M–L bonds somehow?

Showing that this formula works in general is too big a task for this book, but it is possible to see how the maths is consistent with a couple of specific examples.

First, the $Fe(CO)_3$ fragment. By considering the way that a C_{3v} ligand set interacts with the 3d and 4s orbitals of the iron centre, it is possible to derive an MO scheme for the isolated fragment:

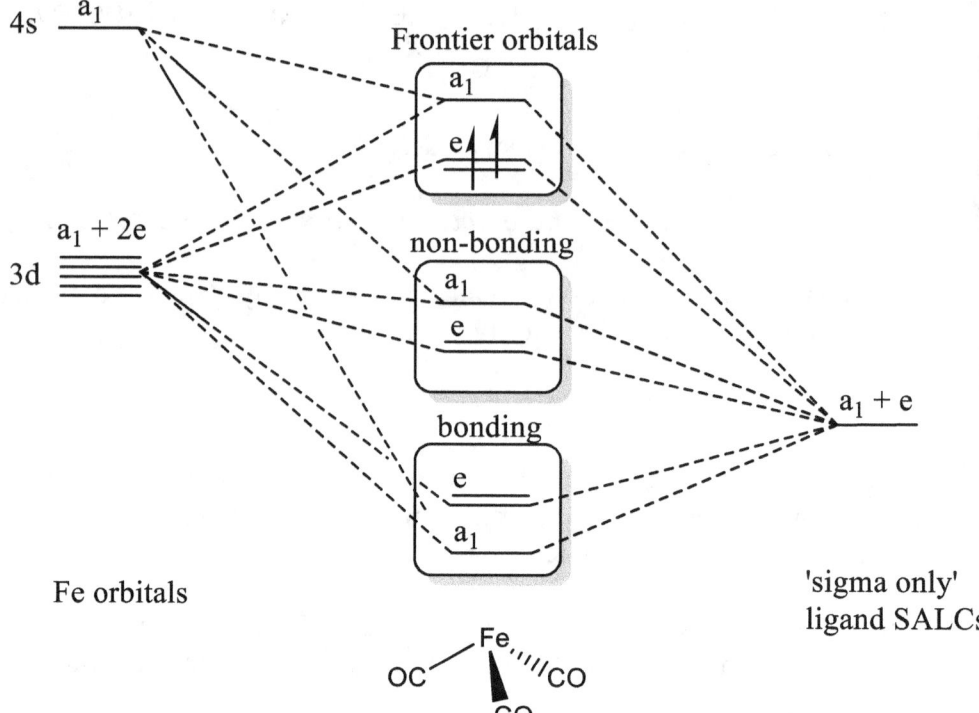

Fe orbitals

Frontier orbitals

non-bonding

bonding

'sigma only' ligand SALCs

6 ligand electrons and 8 metal electrons suggest a ground state with 2 electrons in the antibonding e set (these orbitals have significant d_{xz} and d_{yz} character). Below the HOMO, note that 12 electrons fill the bonding (M–C) and non-bonding orbitals. This is the 12 from the formula. The two surplus electrons (after the bonding and non-bonding MOs in the fragment have been filled) can become involved in cluster bonding.

Similarly, the CoCp fragment can be constructed by considering the Cp π-system as a set of SALCs in C_{5v} symmetry:

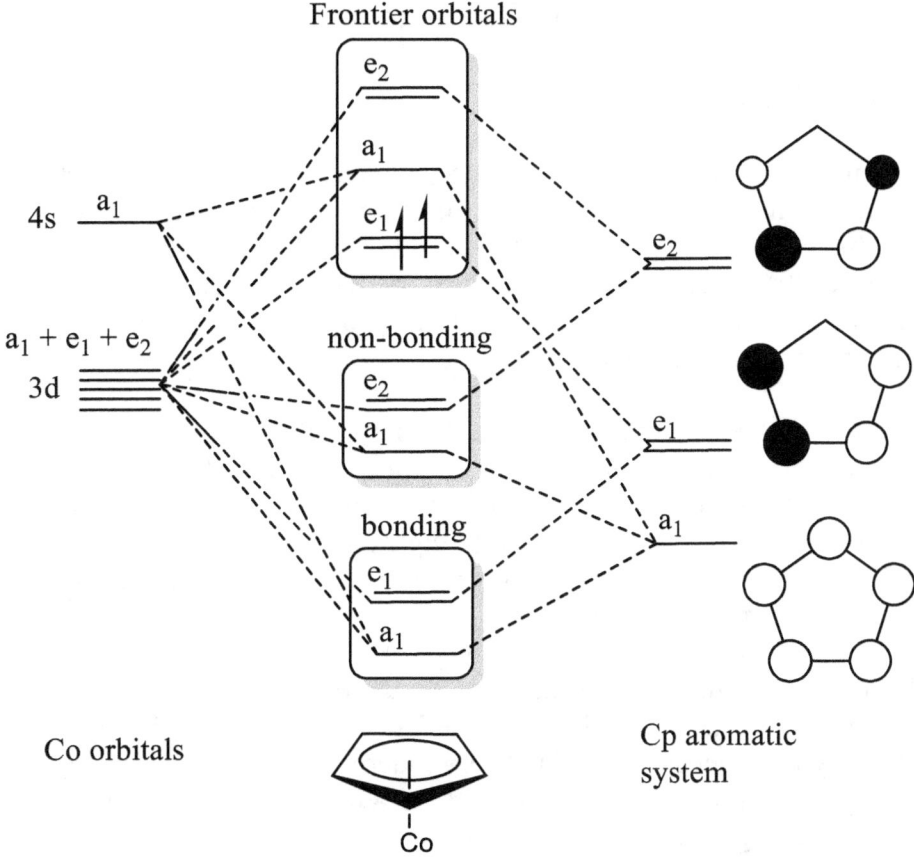

Frontier orbitals

non-bonding

bonding

Co orbitals

Cp aromatic system

Filling all the bonding and non-bonding orbitals requires 12 electrons, leaving only two of the 14 electrons free to bond with the cluster skeleton. In Appendix III the analogy between the boxed 'frontier' orbitals and orbitals in main group fragments such as BH will be drawn explicitly.

Appendix III: The Isolobal Principle

<u>Historical Note</u>

When cluster chemistry was in its ascendance, the dominant model of bonding was orbital hybridisation; this model is still cheerfully used by organic Chemists (sp^3 carbon etc.) despite considerable evidence against it from electronic spectroscopy, perhaps because it allows chemists to think meaningfully about individual atoms in complex asymmetric environments.

In contrast, the hybridisation model has now been mostly abandoned by inorganic chemists in favour of the Linear Combination of Atomic Orbital (LCAO) championed most prominently by Cotton. This approach becomes very intuitive for highly symmetrical systems, but quickly becomes difficult to apply in more complicated structures without computational help.

This two-track history presents a real issue for *teaching* isolobal ideas. The vision presented by chemists like Hoffmann is rooted deeply in hybridisation theory, and is hard for modern inorganic chemists to engage with. I have struggled to find literature 'updating' hybridisation analyses using MO theory.

So, I have represented the orbitals below using a slightly messy mix of hybridisation (for the radial a_1-type sp orbitals) and atomic (for the tangential e-type p-orbitals) models. I hope the pictures carry the discussion: the main point is that analogous orbitals can match in a symmetrically similar way within the cluster even when the fragments are chemically different.

Fragments bond in similar ways if:
1. *They have frontier orbitals of identical symmetry; and*
2. *They have frontier orbitals of comparable energy; and*
3. *These frontier orbitals contribute an identical electron count.*

There are some implicit examples of isolobality in Wade's Rules. For example, the BH fragment can be treated as having frontier orbitals deriving from an sp lone pair and two vacant p-orbitals.

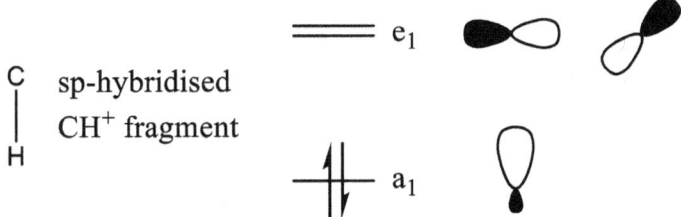

sp-hybridised
BH fragment

This is isolobal with CH^+, which has a very similar orbital scheme (at slightly different energies because of the extra proton in the C nucleus):

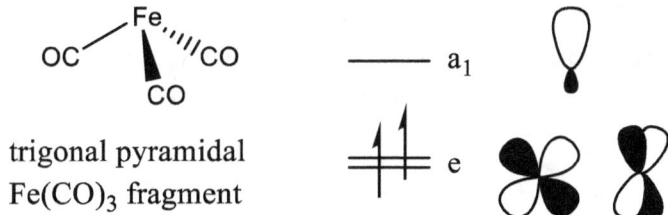

sp-hybridised
CH^+ fragment

But the isolobal principle can be extended further. Consider the frontier orbitals in the $Fe(CO)_3$ fragment:

trigonal pyramidal
$Fe(CO)_3$ fragment

The two electrons in the fragment will sit in the e_1 orbitals rather than the 'lone pair' seen in BH. However, the orbitals can participate in cluster bonding in a similar way to BH. The frontier orbitals are not ordered in the same way, but they have similar-enough energy to join in with the overall cluster bonding scheme: the two-electron $Fe(CO)_3$ is isolobal with BH and CH^+.

Similarly, the two-electron CoCp fragment is part of this isolobal family because of the way its frontier orbitals can develop the same symmetry matches as the BH fragment when it bonds in a cluster.

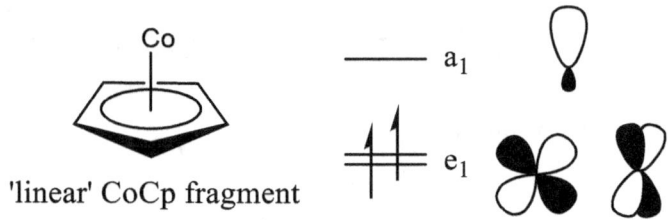

'linear' CoCp fragment

The isolobal relationship is sometimes expressed using a 'looped arrow':

It has never been clear to me whether the loop is supposed to be some kind of orbital, but I suppose it has never been important to me either.

The idea of 'switching' one isolobal fragment for another is a powerful tool for research design because it systematically suggests ways of constructing molecules which haven't been made yet. Designing syntheses to make isolobal analogies actually happen is still very demanding, as the synthetic sources of these synthons are not (in general) interchangeable.

Appendix IV: Selected Deltahedra with Labelling Nomenclature

The numbering for the 'pure' deltahedra is widely-used, but the derivatives of shapes such as the trigonal prism are not normally discussed in this way.

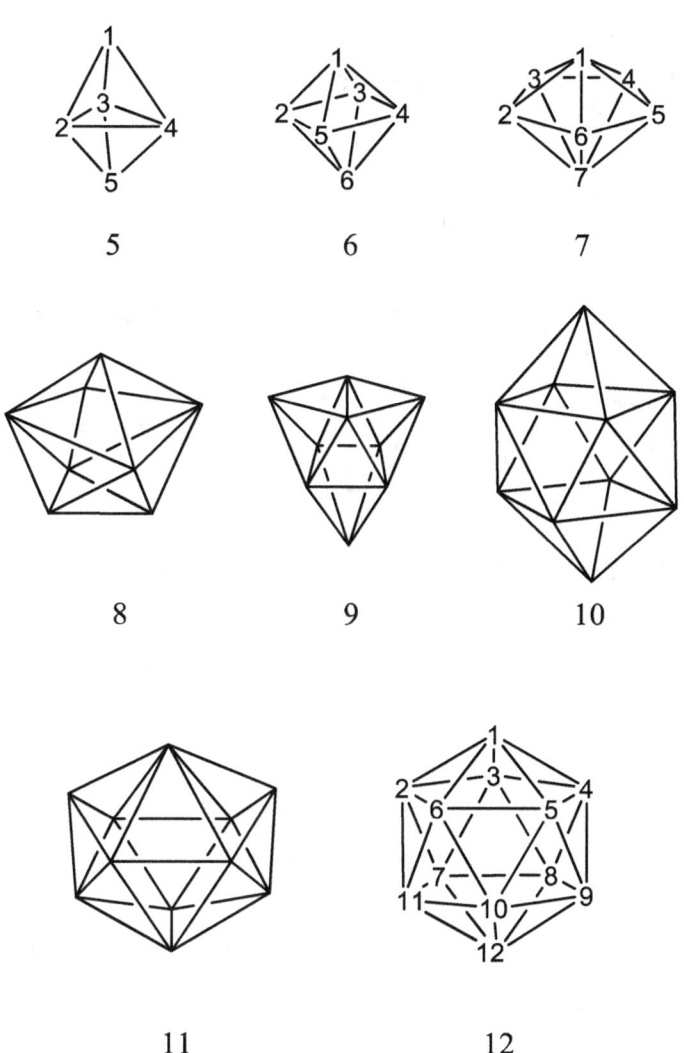

The numbering starts from the rarest type of point (normally apical) and spirals round the molecule (though the chirality of this spiral is arbitrary).

If there is a choice about priority, the priority rules (as used for R/S isomers, disubstituted arenes) should be used; in the first instance this depends upon the atomic number of the atom (i.e. carbon trumps boron).

Note that the term 'basal' means 'on the base' and 'apical' means 'on the apex' of a pyramidal structure.

By this description, B_5H_9 shows four basal BH groups and one apical BH on the 'classic' pyramid.

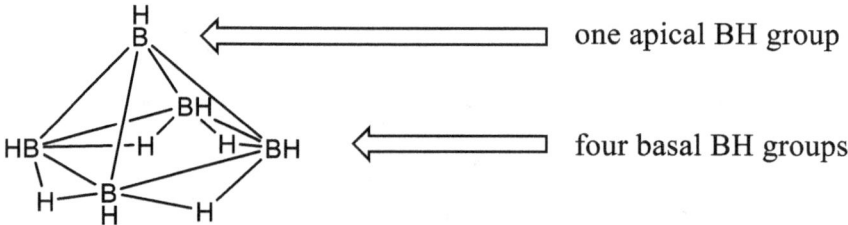

Similarly, in $1,5\text{-}C_2B_3H_5$ the carbon atoms are both in apical positions; the three boron atoms are in the three basal sites.

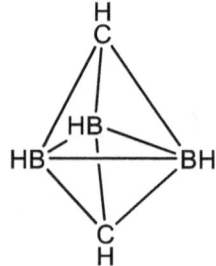

84

1	2	3	4	5	6	7	8	9	10	11	12	13	14	15	16	17	18
H																	He
Li	Be											B	C	N	O	F	Ne
Na	Mg											Al	Si	P	S	Cl	Ar
K	Ca	Sc	Ti	V	Cr	Mn	Fe	Co	Ni	Cu	Zn	Ga	Ge	As	Se	Br	Kr
Rb	Sr	Y	Zr	Nb	Mo	Tc	Ru	Rh	Pd	Ag	Cd	In	Sn	Sb	Te	I	Xe
Cs	Ba		Hf	Ta	W	Re	Os	Ir	Pt	Au	Hg	Tl	Pb	Bi	Po